ENERGY TRANSITION READINESS ASSESSMENT FOR DEVELOPING ASIA AND THE PACIFIC

APRIL 2025

ASIAN DEVELOPMENT BANK

CONTENTS

BOXES

FOREWORD

Developing Asia and the Pacific is crucial for the global energy transition due to its significant population, rapid economic growth, and substantial greenhouse gas emissions, making it essential for achieving global net-zero targets. As the region's economies and populations grow, shifting to sustainable energy systems that deliver on future energy needs is key to achieving national and global climate goals. The transition allows the region to decarbonize its power systems, enhance energy security, improve access to reliable energy for millions, and create jobs while fostering environmental sustainability, economic growth, and climate resilience. Achieving an effective energy transition depends on many enabling factors, including conducive regulations and investment conditions, workforce training, and digital infrastructure.

In light of the urgent need for effective energy transitions in developing Asia, the Asian Development Bank (ADB)—in collaboration with the World Economic Forum—is introducing the energy transition readiness assessment (ETRA) for developing Asia and the Pacific. The ETRA evaluates developing Asian countries' readiness for the energy transition, where "transition readiness" is defined as a country's capacity and preparedness to create an equitable, sustainable, and secure energy system that creates value for society and delivers on net-zero ambitions. The assessment benchmarks countries across critical, independent, and interconnected factors that drive energy transition, offering a multidimensional analysis that accounts for the unique circumstances, capacities, and needs of each country.

Inspired by the World Economic Forum's long-running *Energy Transition Index* (ETI), the ETRA borrows from the ETI's framework, methodology, and indicator selection while building around the unique conditions and needs of developing Asia and the Pacific. The ETRA focuses on ADB developing member countries, representing most of developing Asia's economies, gross domestic product, and population. Complete assessments cover 25 of 41 developing member countries and Brunei Darussalam, with supplementary analysis for small island developing states. By structuring the analysis this way, the ETRA aims to balance comprehensive coverage of diverse countries and indicators while avoiding the analytical hurdles posed by limited data availability in the small island developing states.

The assessment's multidimensional, data-driven, and forward-looking approach ensures its analysis is relevant and effective, respecting each country's pathway to a clean energy future while addressing societal value and sustainability goals. As one-size-fits-all solutions lose effectiveness, tailored pathways for each country are becoming crucial. Recognizing the unique characteristics of the region is essential for effective energy transition strategies. In this context, the ETRA can facilitate regional alignment on climate goals; foster regional collaboration; and serve as a useful diagnostic for policymakers, prospective funders, and investors. The assessment can also bridge the gaps in the knowledge on the implementation of energy transition pathways among countries and encourage greater synergy between countries, aligning them with regional aspirations and global commitments. The region has the potential to leapfrog traditional energy pathways and emerge as a leader in energy transition. With the right strategies and investments, it can set new benchmarks in sustainability and innovation. This assessment aims to catalyze action and collaboration among stakeholders to unlock the potential of developing Asia and the Pacific.

Priyantha Wijayatunga
Senior Sector Director, Energy
Asian Development Bank

Roberto Bocca
Head, Centre for Energy and Materials
Member of the Executive Committee
World Economic Forum

ACKNOWLEDGMENTS

The Energy Transition Readiness Assessment for Developing Asia and the Pacific Region Report 2025 was a collaborative effort led by the Energy Transition Team of the Energy Sector, Sectors Department 1, Asian Development Bank (ADB); and the World Economic Forum, Centre for Energy and Materials.

The report was authored by Annika Seiler, principal energy specialist, ADB; Eric Lam, energy specialist, ADB; Harsh Vijay Singh, head, Equitable Transitions, World Economic Forum; and Maksim Soshkin, research and analysis specialist, World Economic Forum.

Navneeraj Sharma was the data scientist and responsible for the Energy Transition Readiness Assessment (ETRA) design, programming, statistical analysis, and visualization. Maksim Soshkin and Navneeraj Sharma developed the benchmarking tool and codebook for the ETRA, while Ujjwal Bhattacharjee and Kapil Thukral conducted substantial technical editing.

Shant Deyirmenjian, manager, Energy Strategy, Accenture; and Francesca Tate, senior manager, Accenture, provided professional guidance in designing and structuring the quantitative data analysis. Denzel Hankinson, chief executive officer, DHInfrastructure and Kapil Thukral drafted Chapter 4 on Small Island Developing States' energy transition readiness. Virender Kumar Duggal, principal climate change specialist, ADB; Mark Lister, co-founder, Asia Clean Energy Partners; and Bart Lucarelli, managing director, Roleva Energy, contributed special focus pieces to the report.

Eric Lam coordinated the production of this report under the supervision and executive leadership of Pradeep Tharakan, director, Energy Transition, ADB.

Hideaki Iwasaki, director general, Sectors Department 1, ADB; and Priyantha Wijayatunga, senior sector director, Energy, ADB; together with colleagues from the Centre for Energy and Materials, World Economic Forum: Roberto Bocca, head and member of the Executive Committee; Espen Mehlum, head, Energy Transition and Regional Acceleration; and Nicholas Wagner, manager, Energy and Industry Transition Intelligence, provided invaluable strategic guidance.

Special thanks go to Leilei Song and the Economic Research and Development Impact Department in ADB for their review of the ETRA methodology. Thanks as well to Toyo Kawabata, programme officer, International Renewable Energy Agency who coordinated the agency's peer-review of this report.

Melanie Kelleher edited the manuscript. Alvin Tubio typeset and produced the layout. Jan Carlo Dela Cruz created the cover design. Joy Quitazol-Gonzalez proofread the report. Joie Celis assisted in pageproof checking the report, and Mariel Mationg supported in coordination and administrative matters.

Finally, the team would also express their appreciation for ADB's Rafayil Abbasov, senior energy specialist and Sara Afzar, senior partnerships specialist for their invaluable support, without which this report would not have been finalized.

ABBREVIATIONS

ADB	Asian Development Bank
ASEAN	Association of Southeast Asian Nations
CCUS	carbon capture, utilization, and storage
CFPP	coal-fired power plant
CH_4	methane
CO_2	carbon dioxide
DMC	developing member country
EMDE	emerging and developing economy
ETRA	energy transition readiness assessment
FDI	foreign direct investment
GDP	gross domestic product
GHG	greenhouse gas
JETP	just energy transition partnership
kWh	kilowatt-hour
LCT	low-carbon technology
NDC	nationally determined contribution
PRC	People's Republic of China
PV	photovoltaic
R&D	research and development
SIDS	small island developing states
T&D	transmission and distribution

EXECUTIVE SUMMARY

Asia and the Pacific is home to 52% of the world's population, 35% of global gross domestic product, 41% of global energy consumption, and 46% of the world's greenhouse gas emissions. The region's energy mix, which has a 20% higher emission intensity than the global average, is central to achieving global net-zero targets. However, with rising energy demand, reliance on coal power generation, energy import dependence and depth of energy-intensive manufacturing, and hard-to-abate sectors such as aluminum, cement, shipping, and trucking make the region's energy transition even more challenging.

An effective transition to cleaner energy will enhance energy security, mitigate vulnerabilities tied to global energy supply fluctuations, create millions of green jobs, and improve access to reliable electricity for millions of people.

The energy transition readiness assessment (ETRA) framework provides a structured, data-driven, and forward-looking approach to evaluate and analyze the readiness of developing Asian countries for energy transition. The ETRA's relevance stems from the fact that the region's energy transition will be crucial for global efforts to reduce carbon emissions to net zero, and an equitable, secure, and sustainable energy transition is vital to the region's future.

The ETRA analysis covers 25 out of the 41 developing member countries of the Asian Development Bank (ADB) and Brunei Darussalam (collectively termed ETRA countries). The assessment also offers insights to close the existing energy transition gaps for small island developing states (SIDS) (Chapter 4: Energy Transition in Small Island Developing States).

Developed in collaboration with the World Economic Forum, the ETRA takes inspiration from the forum's long-running *Energy Transition Index* (ETI), borrowing from the index's framework, methodology, and indicator selection while ensuring it is built around the unique conditions and needs of developing Asia and the Pacific.

The ETRA framework includes a comprehensive set of dimensions and indicators that reflect the multifaceted nature of energy transition readiness. Countries are scored on a scale of 0 (least ready) to 100 (most ready) across 7 dimensions and 22 subdimensions, further categorized into 61 indicators that help capture specific focus areas. The seven main dimensions include (i) Energy System, (ii) Energy Economy Linkages, (iii) Infrastructure System, (iv) Social System, (v) Technology and Diffusion Ecosystem, (vi) Macroeconomic and Investment Environment, and (vii) Regulatory Environment.

The median scores of each dimension and its indicators are further compared with "advanced economies" and "world" median scores to assess the readiness status of the assessed countries in the larger global context.

ETRA Takeaways

Most developing Asian countries require significant efforts to close the energy transition gaps to enable an effective transition, yet the readiness landscape is varied, indicating distinct opportunities and challenges among countries. Developing Asian countries, while making progress in the pursuit of energy transition, significantly lag behind global levels in five out of seven ETRA dimensions. Most developing Asian countries face challenges in balancing economic growth, energy security, sustainability, and socioeconomic equity. Moreover, the analysis indicates that effective regulations, stable economic climate, favorable investment conditions, and developed infrastructure are correlated, suggesting the opportunity to create a loop that encourages clean energy innovation, adoption, and investment flows. These factors also positively correlate with the social system dimension, suggesting their criticality in ensuring a just energy transition.

There is no one-size-fits-all approach to energy transition. Developing Asian countries must leverage their unique strengths, learn from regional leaders, and address challenges through targeted interventions that recognize the multifaceted nature of transition readiness subject to local conditions. For example, cross-dimensional analysis highlights that some Southeast Asian countries with transition-friendly frameworks can learn from the People's Republic of China's (PRC) ETRA performance to advance their technology and diffusion ecosystem.

Developing Asia's energy transition hinges on a resilient energy system relying on two key pillars: (i) reduced dependence on fossil fuels and (ii) enhanced energy security. The region contributes to nearly half of global greenhouse gas emissions, and its energy sector accounts for three-quarters of the region's greenhouse gas emissions. Coal power generation accounted for 48% of the total energy supply and 57% of the electricity generation in the region in 2022. Phasing out coal power generation by a 2050–2060 timeline is challenging for two reasons. First, coal-fired power plants (CFPPs) in the region are relatively young. Second, many countries rely heavily on coal power generation as coal power continues to be a cheaper resource. The early retirement of CFPPs is possible but would require appropriate carbon pricing, frameworks for transition finance, and substantial amounts of concessional funding while ensuring a just transition for those adversely impacted by coal phaseout. ADB's *Energy Transition Mechanism* initiative, launched in 2021 and piloted in Indonesia, offers a framework for the early retirement of CFPPs anchored around "just-transition" principles.

An energy-secure developing Asia will require a rapid scale-up in renewable energy capacity, which will have cascading effects in lowering the carbon dioxide emissions intensity, reducing net energy imports, diversifying electricity generation, increasing the share of clean energy sources in electricity generation, etc. While more needs to be done, developing Asia is making clear progress toward a massive renewable energy capacity build-out. Of the 26 assessed countries, in 2022, the share of new renewable energy capacity exceeded 75% in nine countries, including the PRC, India, and Kazakhstan.

Balancing economic development and sustainability is key to developing Asia's energy transition aspiration. The region scores below the world and advanced economy medians for the energy economy linkages dimension, implying widespread challenges in separating economic growth from energy consumption and emissions. Many countries need to address inefficient energy production and consumption, high emissions per economic output, and slow adoption of energy-efficient technologies critical for a sustainable transition.

Energy trade in the low-carbon technology products from developing Asia has increased, with the PRC setting benchmarks for the rest of the region. However, the decarbonization of energy-intensive manufacturing must be prioritized by improving energy efficiency, directly and/or indirectly electrifying activities with renewable power, and shifting toward inherently less energy-consuming economic activity.

Developing Asia's energy transition readiness will be determined, in part, by the quality of its energy infrastructure, its ability to handle climate shocks, the flexibility of its power system, and readiness for digital solutions. Overall, developing Asian countries have made progress in reducing transmission and distribution losses from 2000 to 2021. However, the quality of the energy infrastructure still lags considerably compared to advanced economies due to overloaded power transmission and distribution lines and undersized transformers resulting from inadequate planning, unplanned expansion, and poor maintenance practices.

The assessed countries score higher on "flexibility of the energy system" than advanced economies, due to greater integration of hydro and gas power sources within their power infrastructure. For example, Turkmenistan has the most flexible power system among developing Asian countries.

Ensuring a just and equitable energy transition is crucial for developing Asia to sustain socioeconomic progress. Developing Asia has achieved significant milestones in household access to electricity, with more than 1 billion people gaining electricity access in the region since 2010, primarily led by Bangladesh, India, and Indonesia. However, energy and energy services are a significant share of household expenditure for the lower-income population. As the energy transition gathers pace and renewable energy substitutes fossil-fuel power plants, there is potential for short-term shocks in energy prices. Developing Asian countries must design effective energy-related emergency mechanisms to support the affected population.

Developing Asian economies must create an enabling environment to spur innovation in clean energy technology. The technology and diffusion ecosystem dimension scores lowest across all dimensions of the ETRA analysis, with the PRC as an outlier leading in innovating and producing clean energy technologies and services. Beyond the PRC, nine developing Asian economies—including most Southeast Asian states and India—score above the world median for this dimension. Developing Asian countries still face substantial barriers to greater diffusion and production of clean energy products required for the energy transition, especially compared to advanced economies. Generally, low scores in research and development spending, the readiness of frontier technologies, and talent competitiveness suggest a lack of an enabling ecosystem necessary for scale-up.

Developing Asian economies must scale up and broaden private investment in clean energy. Since 2013, clean energy investment in developing Asia has grown more than 900%, reaching $729.4 billion in 2023 and about 45% of global investment. Transition and climate finance are highly concentrated in the PRC, which accounted for nearly 90% of the region's investment during 2013–2023.

Green, sustainability, and sustainability-linked bonds and transition loans are crucial for developing Asian economies. During 2020–2022, the annual issuance of these debt products in developing Asian economies increased from $70 billion to $295 billion, with 89% of the financing coming from the private sector in 2022.

In addition, developing Asian economies must strengthen local financial systems and capital markets. Domestic private investments are vital for financing climate-related projects in local currency reducing reliance on foreign capital. For example, domestic private investment has been a major driver of clean energy investment in larger economies like the PRC and India.

Developing Asian economies must strengthen and harmonize their regulatory frameworks to accelerate a clean energy transition. Developing Asian countries must strengthen energy governance and enforce energy transition policies. Their regulatory quality scores are below the advanced economies' and world median, indicating scope for improvement. Most developing Asian countries have gaps in their climate commitments, with only Bhutan, Georgia, and Sri Lanka on par with advanced economies' median scores for commitment to climate actions.

However, more than half of the assessed countries are progressing in streamlining their regulatory frameworks to support clean energy transition. Georgia, Malaysia, and Thailand are notable performers, scoring close to the advanced economies' median score.

Small island developing states face significant energy transition challenges due to their heavy reliance on imported fossil fuels, economic constraints, limited capacity, and exposure to climate risks. SIDS spend over $1 billion annually on fossil fuels, leaving them vulnerable to global price fluctuations, which may impact their economies and essential services. They are also highly susceptible to climate-related disasters, and energy affordability varies widely, with high tariffs and underdeveloped infrastructure exacerbating the situation. Access to electricity averages 90%, with access to clean cooking at lower rates across SIDS. While renewable energy adoption is critical, barriers such as land-use rights, logistical challenges, and limited local expertise hinder progress. Financial resources for resilient energy infrastructure are scarce, and regulatory frameworks need strengthening to achieve ambitious decarbonization targets. As a result, significant international support, capacity building, and improved regulatory environments will be essential for SIDS to advance their energy transition.

The energy transition for SIDS remains elusive for various reasons, including weak transmission and distribution networks, complexities of the traditional landownership system and limited land, limited manufacturing supply chains, and a lack of a talent pool. Often, the electricity tariffs are below the cost of service, leaving little room for investments in upgrading power systems and building climate-resilient power infrastructure.

Given their vulnerability to climate change impacts, most SIDS have set ambitious decarbonization targets. However, their regulatory apparatus must be strengthened to realize these goals.

1

Background and Regional Energy Context

Developing Asia and the Pacific plays a significant role in the global energy transition. With 52% of the global population and 35% of gross domestic product (GDP) in purchasing power parity terms,[1] the region accounts for 41% of global energy consumption and 46% of greenhouse gas (GHG) emissions, with an energy mix 20% more emission-intensive than the global average.[2] Given its outlook of sustained economic growth (annualized 6.1% GDP growth projected until 2028) and demographic trends, its demand for energy and contribution to global emissions are expected to climb.[3] As a result, the region's energy policies and actions will heavily impact its prosperity and contribution to combating global climate change.

However, an energy transition that balances equity, sustainability, and security for developing Asian countries is fraught with challenges. Fossil fuels remain the backbone of the energy sector, with coal providing about half of the region's primary energy supply in 2020, followed by crude oil (20%) and natural gas (10%) (Figure 1). As a result, coal alone accounts for 70% of energy-related emissions, followed by oil and gas, a structure that has remained largely unchanged since the 1990s. The growth of hard-to-abate industries, such as steel, cement, chemicals, and heavy transport, further complicates decarbonization.

[1]   Population statistics from the UN Population Division Data Portal and the GDP data taken from the International Monetary Fund (IMF) April 2024 World Economic Outlook Database (both accessed 18 July 2024).

[2]   Energy Consumption data from the US Energy Information Administration (accessed 22 October 2024) and the GHG data from the European Commission's Emissions Database for Global Atmospheric Research (EDGAR) Database (accessed 11 July 2024).

[3]   IMF. April 2024 World Economic Outlook Database (accessed 18 July 2024).

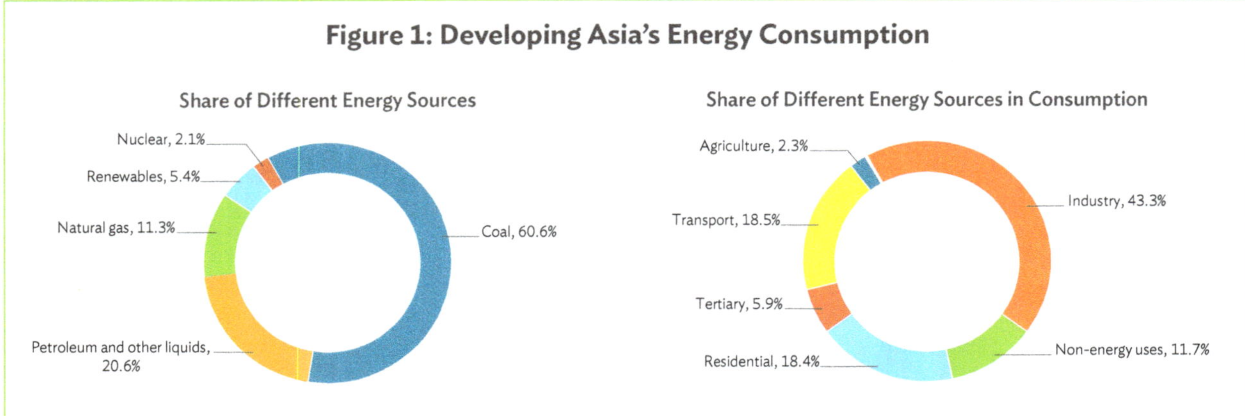

Figure 1: Developing Asia's Energy Consumption

Note: Percentages may not total 100% because of rounding.

Sources: Asian Development Bank based on Government of the United States, Energy Information Administration. International Energy Sources (accessed 22 October 2024), and Enerdata. Energy Balances Data (accessed 2 July 2024).

Despite the growing interest in clean energy technologies, the lack of commercially viable solutions for these sectors poses significant barriers. Increased global geopolitical instability has also added pressure to the region's energy supplies, with most of the countries in developing Asia and the Pacific being net importers of energy.

Despite these obstacles, the energy transition presents immense opportunities. Clean energy investments in the region have grown rapidly, reaching $729.4 billion by 2023. However, aligning with accelerated global net-zero targets will require an estimated $1.4 trillion in annual investments by 2030.[4] Investments must focus on decarbonizing power systems, scaling up renewable energy capacity, and addressing hard-to-abate sector energy needs. The transition will also require significant investment in infrastructure. Between 2020 and 2050, $282 billion annually will be needed for transmission, distribution, and storage systems to support increased renewable energy generation.[5] Emerging technologies, such as hydrogen, advanced batteries, and smart grids, offer promising solutions to modernize energy systems while addressing gaps in access and efficiency. Notably, more than 350 million people in the region do not have an adequate supply of electricity, and 150 million have no access.

Global commitments, including the Paris Agreement and the 29th annual United Nations Climate Change Conference (COP29) initiatives, also provide strong momentum. Under the Paris Agreement, all countries in the region have committed to decarbonizing, and the largest emitters have pledged to reach net-zero emissions before or by mid-century. In 2025, the next round of nationally determined contributions (NDCs) presents an opportunity for countries within the region to develop effective medium-term energy transition strategies underpinned by supportive policies and voluntary commitments including adopting the COP29 pledges to triple renewable energy capacity, double annual energy efficiency improvement rates by 2030, and transition away from fossil fuels equitably, as agreed in the UAE Consensus.

---

[4]   C. Montague, K. Raiser, and M. Lee. 2024. Bridging the Clean Energy Investment Gap: Cost of Capital in the Transition to Net-Zero Emissions. *OECD Environment Working Papers.* No. 245. Organisation for Economic Co-operation and Development.

[5]   Asian Development Bank (ADB). 2023. *Asia in the Global Transition to Net Zero: Asian Development Outlook 2023 Thematic Report.*

Policy measures such as energy subsidy reforms, carbon pricing, and carbon markets are also proving to be strong market signals in the region, driving investment in clean energy technologies.

Decarbonization will also drive job creation. Ambitious climate action could generate 1.5 million additional energy jobs by 2050, primarily in solar photovoltaic (PV) and wind energy industries (footnote 5). However, a just transition is critical. Governments must ensure strong social protections, labor rights, and reskilling programs to support workers affected by the phaseout of fossil fuels. Social support systems must be transformed to balance economic growth with environmental sustainability and equity.

The energy transition is central to the region's climate and development goals. By embracing innovative technologies, scaling up investments, and committing to equitable policies, developing Asia and the Pacific can lead the global fight against climate change while fostering sustainable growth and energy security.

2

The Energy Transition Readiness Assessment

The energy transition readiness assessment (ETRA) framework provides a structured, data-driven, and forward-looking approach to evaluate and analyze the readiness of developing Asian countries for energy transition. "Transition readiness" refers to a country's capacity and preparedness to create an equitable, sustainable, and secure energy system that creates value for society and delivers on net-zero ambitions. To capture this, the assessment benchmarks countries across critical dimensions that measure independent and interconnected factors influencing a country's energy transition readiness. The final result is a multidimensional analysis that considers the unique circumstances, capacities, and needs of each country. This approach allows for a tailored assessment that respects the pathways each country must take toward a clean energy future, ensuring that the strategies are relevant and effective in their specific contexts.

Developed in collaboration with the World Economic Forum, the ETRA takes inspiration from the Forum's long-running Energy Transition Index, borrowing from the index's framework, methodology, and indicator selection while also ensuring it is built around the unique conditions and needs of developing Asia and the Pacific.

The ETRA can facilitate regional alignment with climate goals, foster regional collaboration, and serve as a useful diagnostic for prospective funders and investors. The assessment can also fill knowledge gaps on the implementation of energy transition pathways among countries and encourage greater synergy between countries, aligning them with regional aspirations and global commitments.

Objectives of the ETRA:

- Establish a targeted approach to the specific challenges of developing Asia and the Pacific, guided by a tailored methodology and parameters that resonate with the needs, contexts, and characteristics of the countries, with country examples as case study solutions.

- Facilitate shared regional narratives around an equitable, secure, and sustainable energy transition.

- Support governments in formulating and implementing tailored energy transition strategies in alignment with regional objectives.

- Provide a transparent assessment of the regional energy transition landscape to facilitate investments in clean energy.

- Facilitate knowledge and skills transfer, fostering best-practices exchanges among peer economies in developing Asia and the Pacific.

The ETRA focuses on developing countries because

- **Developing the energy transition of the Asia and Pacific region will be crucial for global efforts to reduce carbon emissions to net zero.**

  - Developing Asia and the Pacific accounted for 52% of the global population and 35% of GDP in 2022, as well as 41% of world energy consumption and about 46% of GHG emissions (footnotes 1–3).

  - The region's GDP is expected to grow at 1.5 times the rate of the global economy until 2028, making it a prime driver of future energy demand and emission output.

  - The region's continued economic growth and industrialization make it home to a growing share of hard-to-abate transport (aviation, trucking, and shipping) and manufacturing (aluminum, steel, chemicals, and cement).

  - The region's economies are important to the global clean energy supply chain.

- **An equitable, secure, and sustainable energy transition is vital to the future of developing Asia and the Pacific.**

  - The majority of economies in the region import more energy than they export and are more energy-intensive relative to the size of their economies, making them more vulnerable to global energy supply shocks. Out of 26 assessed economies, 18 are net energy importers.

  - Access to affordable energy and quality infrastructure is often limited; about 350 million people across the region do not have a reliable electricity supply.[6]

  - Many of the region's countries are also vulnerable to climate change-induced shocks, with about 70% of the global population susceptible to rising sea levels.

  - An effective energy transition can catalyze economic prosperity, with more than 17 million jobs expected to be generated by accelerated decarbonization efforts in developing Asia (footnote 5).

  - An effective energy transition offers vast potential for innovation in clean energy, spur manufacturing, and cascade a multiplier effect across different sectors.

---

[6]   ADB. ADB's Work in the Energy Sector.

While the ETRA aims to provide insights relevant to all developing countries in Asia and the Pacific, it predominantly covers the developing member countries (DMCs) of the Asian Development Bank (ADB). Assessed countries are highly representative of developing Asia, accounting for most of the region's GDP, and population. A full assessment is provided for 25 out of the 41 ADB DMCs and Brunei Darussalam (collectively termed ETRA countries, and listed in the succeeding table).[7] Chapter 4 (Focus on Small Island Developing States) offers a supplementary analysis of small island developing states not covered in the full assessment due to limited indicator data coverage.

Energy Transition Readiness Assessment Country Coverage

With Full Assessment		With Partial Assessment
Afghanistan	Myanmar	Cook Islands
Armenia	Nepal	Fiji
Azerbaijan	Pakistan	Kiribati
Bangladesh	Philippines	Maldives
Bhutan	Sri Lanka	Marshall Islands
Brunei Darussalam	Tajikistan	Micronesia, Federated States of
Cambodia	Thailand	Nauru
China, People's Republic of	Turkmenistan	Niue
Georgia	Uzbekistan	Palau
India	Viet Nam	Papua New Guinea
Indonesia		Samoa
Kazakhstan		Solomon Islands
Kyrgyz Republic		Timor-Leste
Lao People's Democratic Republic		Tonga
Malaysia		Tuvalu
Mongolia		Vanuatu

ETRA = energy transition readiness assessment.

Notes: Full assessment includes analysis based on the ETRA's 61 indicators and 7 dimensions. Partially assessed countries are not covered in this analysis but are included in Chapter 4: Focus on Small Island Developing States.

Source: ADB Classification.

The ETRA report should be read in conjunction with the country-by-country analysis. Additional technical background is provided in stand-alone documents, including a detailed methodology and codebook.

---

[7]   Brunei Darussalam is grouped along with 25 ADB DMCs in the ETRA analysis. Brunei Darussalam faces challenges and opportunities in its pursuit of energy transition readiness similar to the 25 DMCs.

Energy Transition Readiness Assessment Framework

The ETRA framework is designed to assess countries based on their readiness for the energy transition, encompassing economic, technical, and risk-related factors. It includes a comprehensive set of indicators and dimensions that reflect the multifaceted nature of energy transition readiness. Countries are scored on a scale of 0 (least ready) to 100 (most ready) across seven dimensions, which are further categorized into 22 subdimensions, which comprise 61 indicators (Figure 2) that help capture specific areas of focus.

Energy System: Assesses a country's capability to ensure energy security and transition away from fossil fuels at speed and scale. Scaling up low-carbon energy sources while transitioning from fossil fuels depends on the extent of current carbon lock-in, resource endowments, and system architecture. This dimension measures the ability of countries to diversify energy imports and supply, integrate clean energy sources, improve emission and energy intensities, and reduce fossil fuel dependence.

Energy Economy Linkages: Assesses a country's capability to decouple energy use from economic growth and development. Energy is interlinked with economic growth, impacting industrial production, export revenue, and fiscal stability. This dimension assesses the ability of countries to adopt best-in-class energy-efficient technologies to improve both the emission and energy intensities of its economy, increase trade in low-carbon technology (LCT) products, and reduce the carbon footprint of its exports.

Infrastructure System: Assesses a country's capability to integrate low-carbon energy sources, optimize energy demand, and improve transportation systems and development of information and communication technology. It measures countries' climate resilience and coping capacity, energy infrastructure quality, and physical and digital infrastructure.

Social System: Assesses a country's capability to provide affordable energy and services and to recognize and address the distributional impacts of the energy transition. This dimension considers select key parameters to assess countries' capabilities to ensure access to affordable energy and energy services, rationalized electricity tariffs for vulnerable sections of the population, and institutional mechanisms to identify and address distributional implications of the energy transition on vulnerable stakeholder groups.

Technology and Diffusion Ecosystem: Assesses a country's capability to innovate, transfer, produce, and deploy new energy technologies. It measures a country's research and development (R&D) ecosystem, availability of skilled human capital, clean energy technology diffusion, and comparative advantage.

Macroeconomic and Investment Environment: This dimension assesses a country's capability to attract diverse sources of domestic and foreign capital and its macroeconomic fundamentals. It focuses on monetary and fiscal conditions, financial stability and risk, openness to investors, and clean energy investment flows that influence and reflect its ability to attract and allocate funding for the energy transition.

Regulatory Environment: Assesses a country's ability to create stable, comprehensive energy transition policies. Effective regulations align stakeholder interests, enhance governance, support modern infrastructure and innovation, ensure energy security and resilience, guide sustainable investments, minimize macroeconomic risks, and promote social equity by ensuring affordable energy and a just transition.

Figure 2: Energy Transition Readiness Assessment Dimensions, Subdimensions, and Indicators

Energy System

Energy Security
- Diversity of Electricity Generation
- Diversity in Energy Supply
- Diversification of Energy Imports
- Net Energy Imports

Clean Electrification
- Share of electricity in final energy consumption
- Share of clean energy sources in electricity generation

Fossil Fuel Dependence
- Average Life of Coal Plants
- CO_2 Intensity of TPES
- CH_4 Intensity of Production

Energy Economy Linkages

Energy-Emissions-Economy Nexus
- Energy Intensity
- Emissions Intensity

Energy Trade
- Emissions Intensity of Exports
- Trade Balance in LCT

Resource Dependence
- Net Fuel Imports
- Fossil Fuel Rents
- Energy Subsidies

Infrastructure System

Climate Resilience and Coping Capacity
- Water Stress
- Lack of Coping Capacity
- Infrastructure Vulnerability

Energy Infrastructure Quality
- Transmission and Distribution Losses
- Flexibility in Electricity System

Physical and Digital Infrastructure
- Quality of Infrastructure
- Digital Infrastructure Readiness

Social System

Energy and Transportation Access
- Per Capita Energy Consumption
- Rural Electricity Access
- Clean Cooking and Heating Access
- Access to Public Transportation

Energy Affordability
- Electricity Prices for Industry
- Household Electricity Prices
- Petrol Prices

Social Equity and Development
- Human Development Index
- Social Protection Cover
- Labor Rights

Environmental Health and Climate Vulnerability
- PM2.5 Air Pollution, Mean Annual Exposure (micrograms per cubic meter)
- Vulnerability to Climate Disruptions

Technology and Diffusion Ecosystem

Clean Energy Technology Development and Innovation Capacity
- Public R&D
- Frontier Technologies Readiness

Human Capital
- Talent Competitiveness
- Jobs in Low Carbon Industries

Technology Diffusion
- Renewable Capacity Build-out
- Comparative Advantage in LCT

Macroeconomic and Investment Environment

Monetary and Fiscal Indicators
- General Government Debt
- Fiscal Balance
- Consumer Prices Inflation
- Exchange Rate Depreciation
- Real Interest Rates

Financial Stability and Risk
- Domestic Credit to Private Sector
- NPA Share of Gross Bank Loans
- Risk of Banking System Failure
- Sovereign Credit Rating

Investment Climate
- Investment in Clean Energy
- FDI in Clean Energy Sector
- De Jure Financial Globalization

Regulatory Environment

Governance Quality
- Regulatory Quality
- Government Effectiveness
- Statistical Performance

Energy Policy Framework
- Renewable Energy Policies
- Clean Cooking Policies
- Energy Efficiency Policies
- Electricity Access Policies

Climate Commitments
- Climate Action Commitment

CH_4 = methane, CO_2 = carbon dioxide, FDI = foreign direct investment, LCT = low-carbon technology, NDC = nationally determined contribution, NPA = nonperforming asset, PM2.5 = particulate matter, R&D = research and development, TPES = total primary energy supply.

Source: Asian Development Bank.

Wind and solar farm. A technician on his high post is repairing an electrical pole at the Burgos Wind and Solar Farm, Ilocos Norte, Philippines (photo by ADB).

3.1 Cross-Dimension Takeaways

The ETRA framework encompasses seven dimensions and reveals a nuanced landscape of progress, challenges, and opportunities across developing Asian countries (Figure 3).

The ETRA performance indicates that most developing Asian countries must address common readiness gaps for an effective energy transition. Figure 3 shows that median scores of assessed economies are substantially below global levels for five out of seven readiness dimensions, with only a handful of countries scoring above advanced economies in select dimensions. Typically, below global scores for energy systems, energy economy linkages, infrastructure, and social systems dimensions underscore that balancing economic growth, energy security, sustainability, and equity is particularly challenging for most developing Asian countries. Typical issues include fossil fuel dependence, carbon-intensive economic growth, underdeveloped and at-risk infrastructure, social inequity, and limited access to affordable and quality infrastructure. Scores for energy systems across all developing Asian countries are especially close, reflecting common challenges in energy security (including reliance on energy imports) and widespread dependence on coal, among other factors.

Most developing Asian countries score low in the technology and diffusion ecosystem, representing the largest performance gap between developing Asian and advanced economies. This highlights an urgent need to strengthen country capacities for clean energy innovation, adoption, and production, without which they risk falling short of decarbonization goals and missing economic opportunities offered by the energy transition.

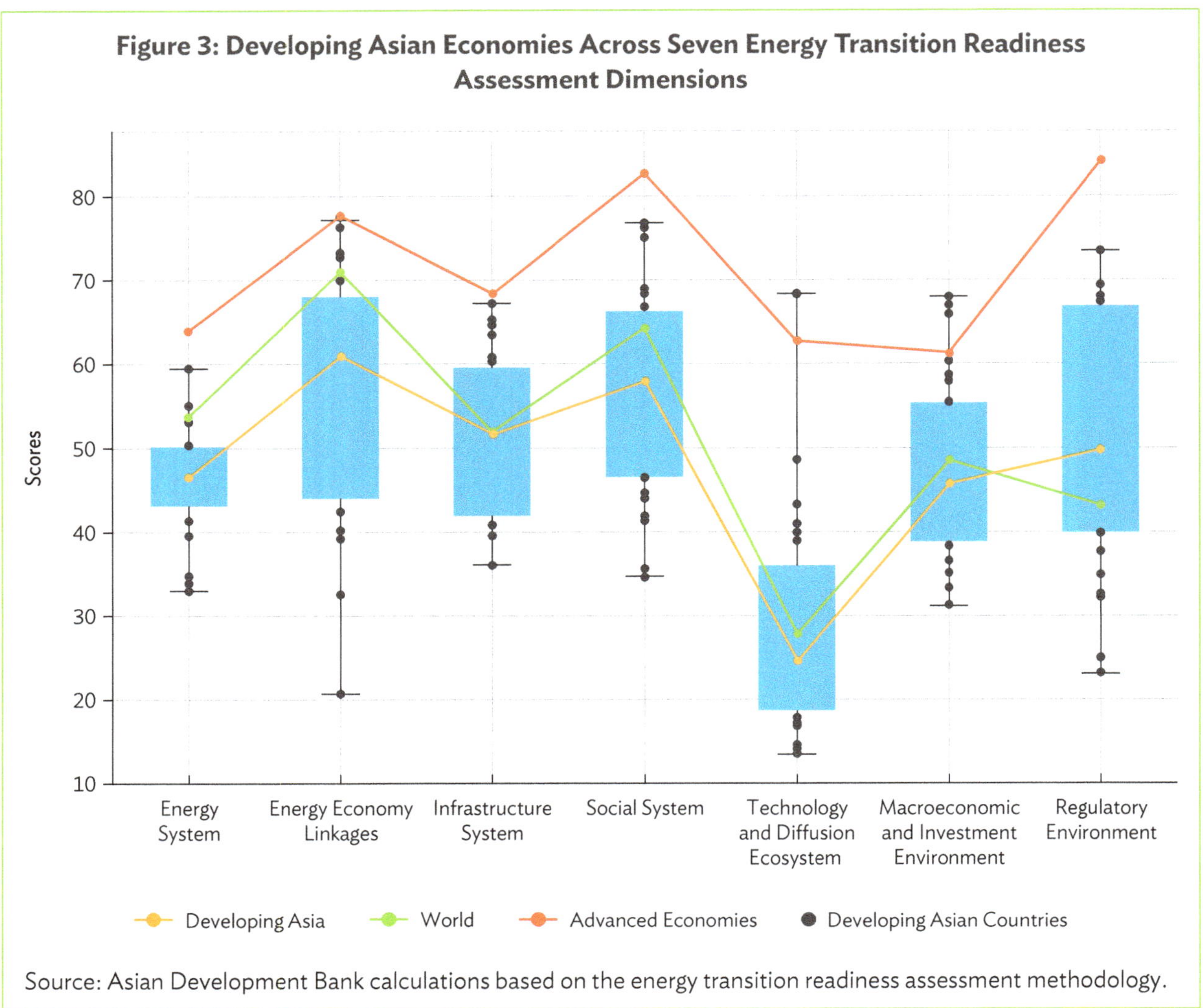

Figure 3: Developing Asian Economies Across Seven Energy Transition Readiness Assessment Dimensions

Source: Asian Development Bank calculations based on the energy transition readiness assessment methodology.

The ETRA analysis reveals the interplay between clean energy innovation and diffusion, investment and infrastructure conditions, and regulatory policy. In particular, the technology and diffusion ecosystem dimension scores correlate with scores for the regulatory environment, macroeconomic and investment environment, and infrastructure system dimensions. **This suggests that the combination of effective regulations, a stable economic climate, favorable investment conditions, and developed infrastructure can create a positive feedback loop that encourages clean energy innovation, adoption, and investment.** Moreover, these factors positively correlate with the social system dimension scores, indicating how these variables are critical for supporting a just energy transition.[8] However, when compared to advanced economies, many developing Asian countries tend to face greater macroeconomic and regulatory uncertainty, underdeveloped energy policy frameworks and infrastructure, higher capital costs, and investment barriers.

---

[8] The regulatory environment, macroeconomic and investment environment, and infrastructure dimensions have a correlation coefficient above 0.7 with the technology and diffusion ecosystems and the social systems dimensions. The only exception is a 0.64 correlation between macroeconomic and investment environment and social systems dimensions. Analysis based on all countries for which data was available.

Moreover, despite commonality, the ETRA analysis indicates that transition readiness can vary greatly between economies and subregions, creating different challenges and opportunities. For instance, the People's Republic of China (PRC) ETRA performance reflects relatively stronger regulatory framework and enabling policy environment, developed infrastructure, and a robust technology and diffusion ecosystem, underscoring the elements that have helped the country become a global leader in clean energy, innovation, adoption, and investment. While typically not on par with advanced economies, ETRA results show that some countries in Southeast Asia are home to more transition-friendly macroeconomic and investment conditions, infrastructure, regulatory landscapes, and technology and diffusion ecosystem. Along with the PRC and India, these economies are best positioned to increase participation in the regional and global clean energy supply chains, with increased clean energy diffusion helping them improve their energy security and position as competitive manufacturers of low-carbon goods. Meanwhile, landlocked countries in Central Asia and the Caucasus tend to score lower for energy economy linkages due to their economic and trade dependence on fossil fuel extraction and energy, making their energy transition more challenging. The diversity of challenges and achievements for each developing Asian country underscores that there is no one-size-fits-all approach to the energy transition. Countries must leverage their unique strengths, learn from regional leaders, and address challenges through targeted interventions that recognize the multifaceted nature of transition readiness subject to local conditions.

The cross-dimensional analysis of the ETRA framework—grounded in the comprehensive data visualized in Figure 3—illuminates the intricately interconnected nature of energy transition readiness. It reveals that developing Asian countries face a complex landscape where progress in one area can have ripple effects across multiple dimensions, often in unexpected ways.

To meet its growing energy needs and decarbonization goals, developing Asia needs to substantially increase public and private investment in clean energy. This requires creating an ecosystem where clean energy integration, efficiency improvements, technological innovation, and social equity harmonize through adaptive regulatory frameworks and resilient infrastructure.

As developing Asian countries navigate this transformative journey, the ETRA framework serves as a crucial compass, offering data-driven insights for benchmarking progress, identifying critical gaps, and informing strategic decisions. By embracing a holistic approach to energy transition—grounded in these cross-dimensional insights—DMCs can meet their climate commitments and unlock new pathways for economic prosperity and social development in a sustainable, low-carbon future.

3.2 Energy System

Developing Asia is at a critical juncture in its journey toward sustainable development. It faces the dual challenge of maintaining robust economic growth while addressing significant environmental concerns, particularly GHG emissions, air pollution, and the destruction of natural capital. Developing Asian countries represent 41% of global energy use and are responsible for nearly half of global carbon emissions, but their per capita energy use and emissions remain low. The ETRA energy system dimension assesses developing Asian countries' preparedness for the clean energy transition in enhancing energy security and reducing dependence on fossil fuels (Figure 4).

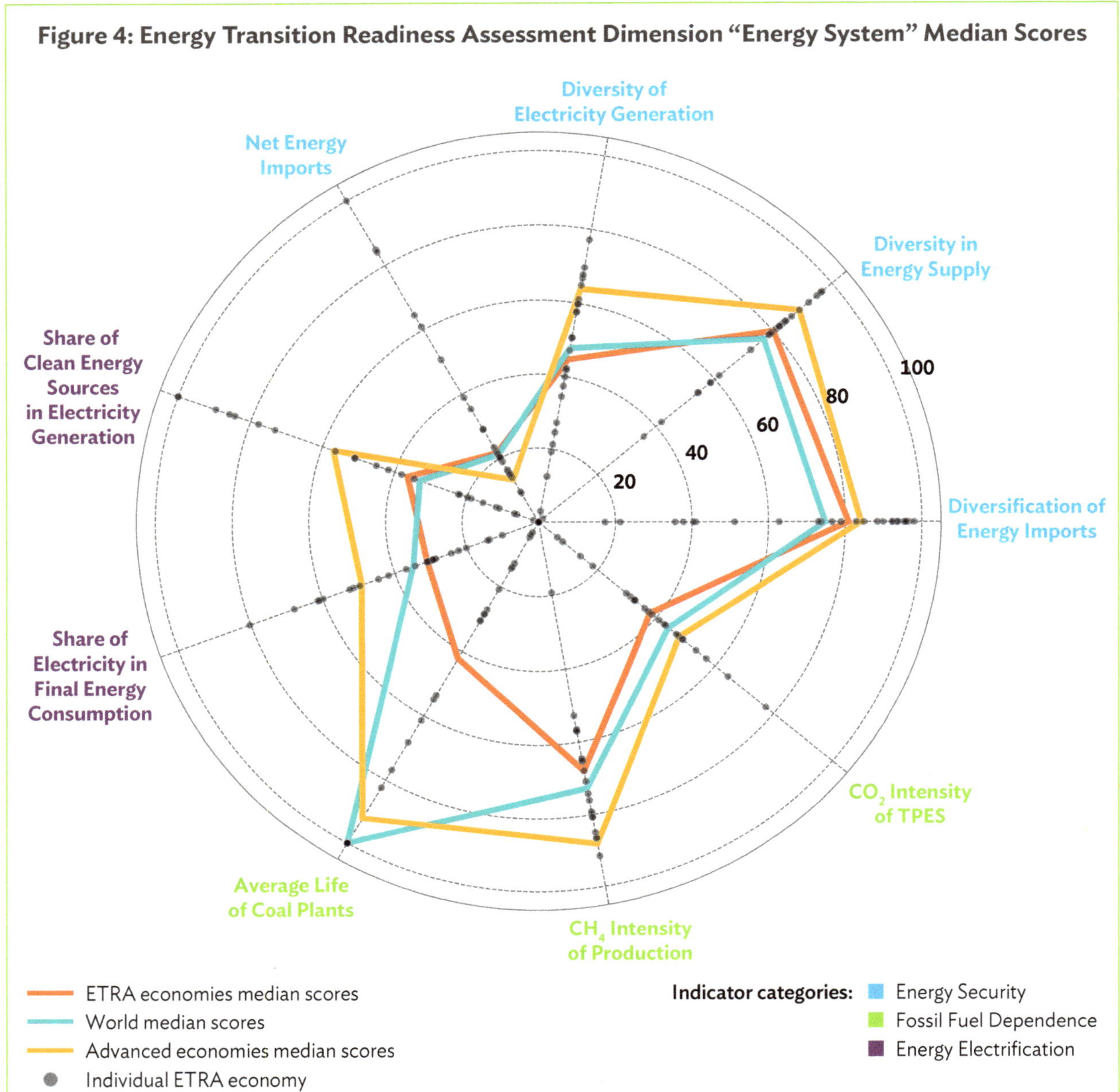

Figure 4: Energy Transition Readiness Assessment Dimension "Energy System" Median Scores

Radar Plot Highlights

- Energy security in developing Asian countries is highly vulnerable to supply chain and climate transition risks. A lower score diversity in energy supply and electricity generation further underscores their high vulnerability to shocks.

- CH_4 and CO_2 emissions intensities are higher among developing Asian countries, demonstrated by a lower median score as the region is highly dependent on coal and gas power generation and lags on adopting clean technologies and energy efficiency.

- A low median score for the average life of coal plants reflects the long average life of coal-fired power plants, making coal-power generation phase-down in developing Asia challenging.

CH_4 = methane, CO_2 = carbon dioxide, ETRA = energy transition readiness assessment, TPES = total primary energy supply.

Note: Lower score = worse, higher score = better.

Source: Asian Development Bank.

Accelerating the clean energy transition is a comp ex challenge for many developing Asian countries, requiring the phaseout of fossil fuels and greater end-use electrification while ensuring energy systems remain affordable, reliable, and resilient to climate risks and shocks. The developing Asian countries' median energy system dimension score lags advanced economies' by about 27%. However, the region is on par with the world median score. Developing Asia's challenge in the energy transition is to meet the rapidly rising domestic energy demand of their growing economies securely, affordably, and sustainably, which continues to be governments' key energy policy priority. The readiness to meet this challenge depends on the resilience of energy security, the extent of current carbon lock-in, resource endowments, and system architecture, all measured in the energy system dimension.

Developing Asia must prioritize diversifying its energy supply to ensure energy security in the short, medium, and long term. The assessed countries' lower median score in the diversity of energy supply compared to advanced economies indicates relatively higher energy security risks. Lower middle-income countries are particularly vulnerable to supply chain bottlenecks, trade disruptions, and geopolitical tensions that may negatively impact the availability and affordability of energy imports. The Russian invasion of Ukraine in February 2022 exemplified this, as it triggered significant disruptions in global energy markets and led to severe energy supply bottlenecks, diverting energy resources away from developing Asia. This rendered gas unaffordable and made it difficult to obtain supply for some assessed countries such as Bangladesh.[9]

Yet, the region's large net energy importers—including the PRC, India, Thailand, and Viet Nam—have taken measures to diversify the geographies of their energy imports, with their individual net-energy import scores significantly higher than advanced economies' median score.

Fuel-exporting developing Asian countries must create long-term strategic plans to diversify economies, and seize opportunities from the global energy transition. The region is also home to some of the world's largest coal and natural gas fuel exporters, including Indonesia, Kazakhstan, and Mongolia. These countries must prepare their economies against climate transition risks from decreasing export markets and stranded assets in their logistics value chain that support the export of their fossil fuel reserves. Many of these countries can leverage their large deposits of mineral and metal resources used in the manufacturing of clean energy technologies to diversify their economies, while also catalyzing green industrialization and equitable development.

Developing Asia must also diversify its energy imports and domestic fuel mix to achieve future electrification of energy end use and reduce vulnerability to supply chain and climate transition risks. While developing Asian countries maintain a median score slightly above the world score for diversity in energy supply and diversification of energy imports, they score lowest in the diversity of electricity generation, reflecting the region's reliance on similar power generation sources.

This reliance, in the medium-to-long term, exposes the countries to high climate transition and energy security risks. For example, fossil fuel-rich economies such as Azerbaijan and Uzbekistan are already experiencing an increasing depletion of domestic gas resources, requiring them to import natural gas to meet their energy demand. On the other hand, countries reliant on hydropower for power generation also face energy security risks as climate change threatens the long-term availability of hydropower resources.

---

[9]   S. Bhowmick. 2022. Demand-Supply Dynamics in Bangladesh's Energy Sector. *Observer Research Foundation.* 29 December.

Bhutan, Georgia, the Kyrgyz Republic, the Lao People's Democratic Republic, Nepal, Pakistan, and Tajikistan are all countries that rely on hydropower generation and face energy security risks as the impacts of climate change arise. Efforts to diversify domestic fuel sources and energy imports can reduce energy security risks and support efforts to transition to cleaner electricity end use.

Developing Asia and the Pacific must accelerate efforts to decarbonize and reduce their energy sectors emission intensity. The region includes some of the world's most carbon and resource-intensive countries. In 2023, the share of the region's global GHG emissions reached 47%. Energy-related CO_2 emissions have increased 1.5 times since 2000, with coal accounting for 48% of total energy supply, and 57% of the region's electricity generation in 2022.[10]

Developing Asia's energy sector accounts for three-quarters of the region's total GHG emissions. Across all sectors, electricity and heat production are the largest and fastest-growing sources— contributing about 40% of emissions—followed by manufacturing (18%); agriculture, land use change, and forestry (13%); and transport (about 6%).[11] Figure 5 illustrates the intensities of CO_2 and CH_4 emissions in developing Asian economies' energy systems. It demonstrates that developing Asian economies remain heavily reliant on coal power generation, resulting in CO_2 intensity continuously increasing since 2012.

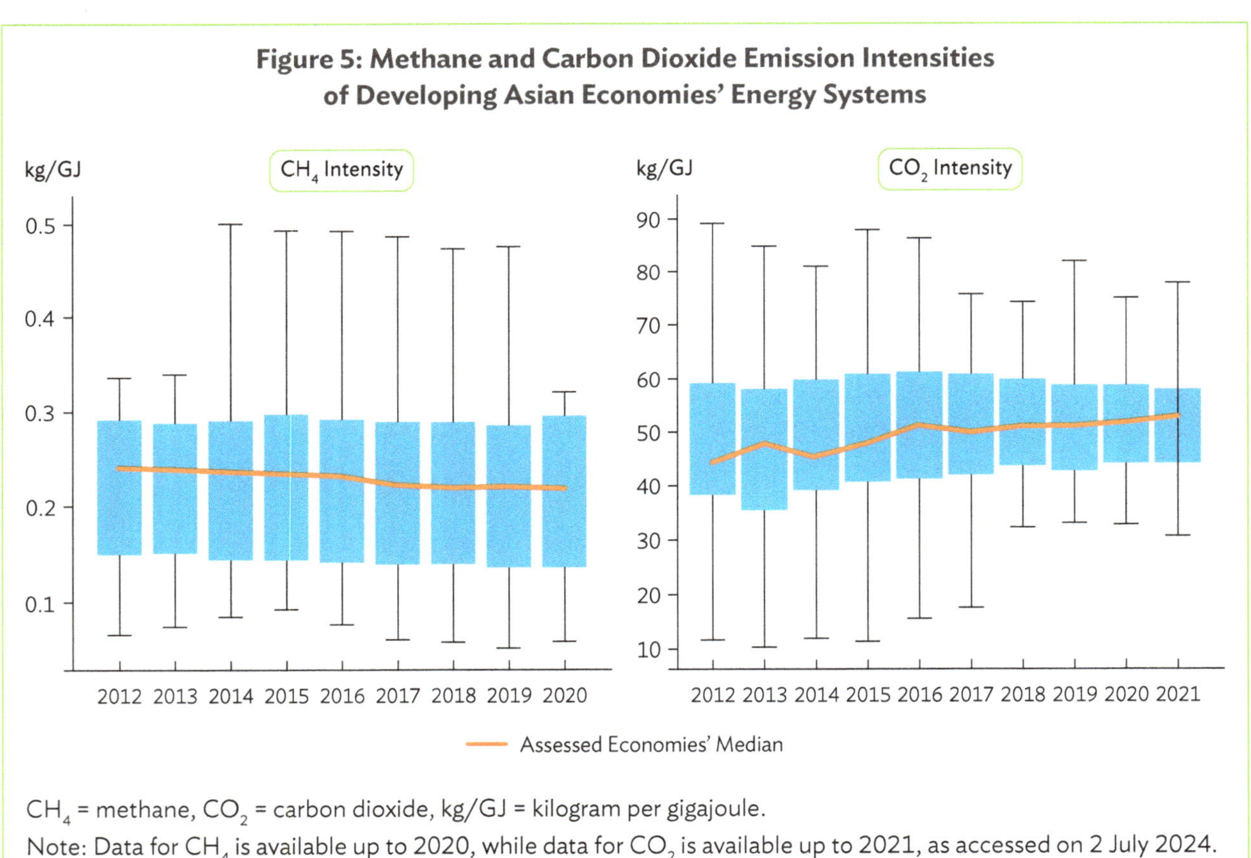

CH₄ = methane, CO₂ = carbon dioxide, kg/GJ = kilogram per gigajoule.
Note: Data for CH₄ is available up to 2020, while data for CO₂ is available up to 2021, as accessed on 2 July 2024.
Source: International Energy Agency GHG Emissions Statistics (accessed 2 July 2024).

---

[10]   Footnote 5, pp. 33–34.

[11]   Footnote 5, p. 15.

Comparatively, developing Asian economies energy systems CH_4 intensity has decreased since 2012, indicating potential of fuel switching either toward cleaner technologies or more carbon-intensive ones. Under the policies and assessed countries' resource-intensive economic model, the expected rapid population and economic growth will drive an unsustainable increase in carbon and methane emissions, placing excessive strains on the environment and biodiversity of ecosystems. This growth trajectory will not likely materialize without strong attention to sustainable growth and require collaborative efforts to reduce emissions (Box 1).

Box 1

Methane Mitigation Efforts in Developing Asia

Developing Asian countries are implementing targeted methane mitigation initiatives to align with global climate goals and enhance sustainability. The Global Methane Pledge, launched at the 2021 United Nations Climate Change Conference (COP26), unites 159 countries, including Bangladesh, Indonesia, Malaysia, Pakistan, the Philippines, and Viet Nam, to reduce methane emissions by 30% from the 2020 levels by 2030.[a]

Kazakhstan joined the Global Methane Pledge in 2023, collaborating with the United States to introduce national standards for leak detection and eliminate non-emergency methane venting in oil and gas by 2030 and plans to invest at least $1.4 billion to limit methane emissions.[b]

Member countries of the Association of Southeast Asian Nations are engaging in methane reduction projects, supported by international collaborations and technology-sharing including methane monitoring to support agricultural resilience and improvements in air quality.[c]

---

[a] Global Methane Pledge.

[b] US Department of State. 2023. US–Kazakhstan Joint Statement on Accelerating Methane Mitigation to Achieve the Global Methane Pledge. Press release. 2 December.

[c] Petronas. 2024. PETRONAS and Partners to Advance Methane Emissions Reduction Efforts in Southeast Asia Region. News release. 2 October.

Assessed economies must prioritize decarbonizing their power systems through an accelerated renewables build-out coupled with broad-based measures to implement smart and flexible grid technologies and energy storage solutions. At the same time, countries must spur energy efficiency in their economies and promote decarbonization of their transport and hard-to-abate industrial sectors. The diversification of assessed countries' energy mix can continue to support the electrification of energy end use within these energy system sectors.

Developing Asian countries must strive to electrify their energy end use. The median score of assessed countries in their share of electricity in final energy consumption was nearly 40% below advanced economies and 11% below world median scores. As the region sustains its development trajectory, the imperative to decarbonize economies through electrification has never been more urgent. This is particularly true given the advancements in renewable energy and energy storage technologies, which present unprecedented opportunities to transform energy systems. From 2000 to 2021, developing Asia's electricity consumption per capita increased by 168%, reflecting rapid economic growth and regional progress in energy access and supply. To reach net-zero by 2050, 90% of total primary energy supply will need to be electrified and electricity must come from low-carbon sources.[12] Since 2010, the average share of electricity in energy end use in developing Asian economies has increased by 5% compared to 3% globally.[13] Over one-third of assessed countries' electrification of energy end use exceeds the world median score. However, two-thirds are still lagging.[14]

With an expected rapid increase in the electrification of energy end use, fuel mix diversity for electricity generation is set to become a key priority for the energy security of the ETRA countries' energy systems. Assessed country median scores underperform the median score of advanced economies for the diversity of the power generation fuel mix by about 30%, reflecting the limited interconnectivity of power systems and electricity trade among countries and energy policy focused on self-reliance on domestic sources for electricity supply. Most countries with a high share of fossil fuels in their fuel mix are predominantly resource-rich countries that rely on domestic resources for their electricity supply.

A rapid scale-up of renewable power capacity for developing Asian countries can strengthen their energy security and the sustainability of energy systems. Driven by the PRC, the region tripled its installed renewable energy capacity during 2014–2023.[15] Renewable energy capacity additions of nearly 300 gigawatts by the PRC in 2023 dwarf the progress of any other country, not just in the region but globally. However, other assessed countries are lagging in terms of their share of clean energy. Countries can strengthen their energy security by advancing the build-out of renewable energy resources, provided their clean energy technology supply chains are diverse and resilient. Figure 6 illustrates developing Asian economies' share of renewable energy and new capacity addition in 2022. The median score of assessed countries is slightly above the world score in their share of low-carbon energy sources in electricity generation but lags the score of advanced economies. The assessed countries can be classified into four quadrant groups based on their share of clean energy sources in new power generation capacity installed and their share in electricity generation.

In 2022, renewables—as a share of new power generation capacity installed—exceeded 75% in nine countries, including the PRC, India, and Kazakhstan, illustrating these countries demonstrated capability, willingness, and enabling ecosystem to install renewable energy infrastructure. On the other hand, carbon-intensive economies in Southeast Asia, such as Malaysia, Indonesia, the Philippines, and Thailand, as well as in Central and West Asia—including Azerbaijan, Turkmenistan, and Uzbekistan—have shown the opposite with low adoption of renewable energy generation and capacity. As such, these countries need to set the appropriate policy directions and enabling environments to meet massive investment requirements (Box 2). Developing Asian countries will need to set incentives, align subsidies targeting and financial incentives, and encourage private sector participation to meet massive investment requirement.

---

[12] International Energy Agency (IEA). 2023. *Net Zero Roadmap: A Global Pathway to Keep the 1.5C Goal in Reach— 2023 Update.*

[13] Noteworthy increases in the share of electricity in the final energy consumption include Bangladesh (+11.5 percentage points to 25%), the PRC (+10 percentage points to nearly 30% in 2023), and Viet Nam (+14 percentage points to 28%). Enerdata. World Energy and Climate Statistics 2024 (accessed 2 July 2024).

[14] The nine countries comprise Bangladesh, Brunei Darussalam, the PRC, Georgia, the Kyrgyz Republic, Malaysia, the Philippines, Tajikistan, and Viet Nam.

[15] ADB, Bloomberg Philanthropies, ClimateWorks, and Sustainable Energy for All. 2023. *Renewable Energy Manufacturing: Opportunities for Southeast Asia.*

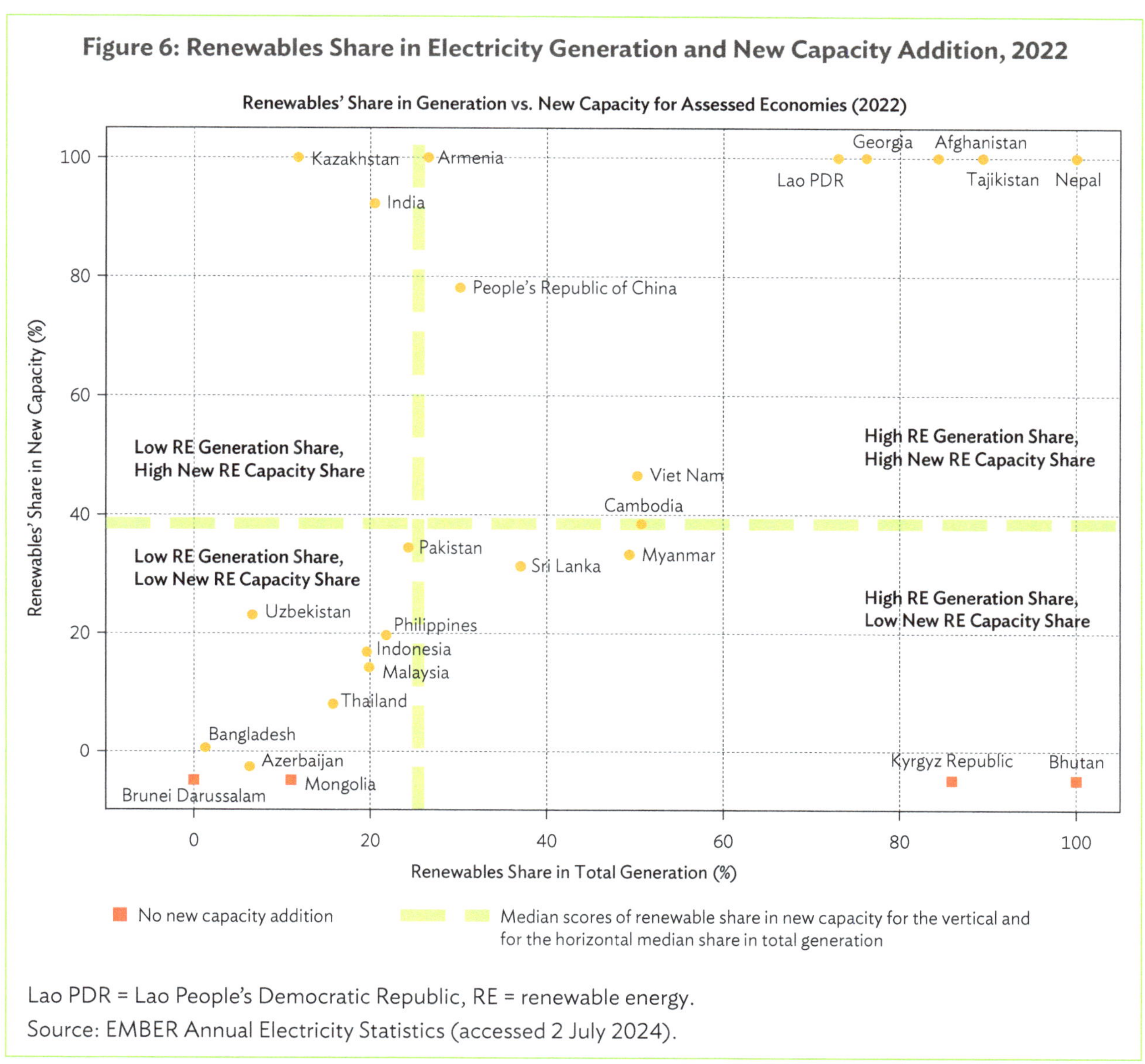

Figure 6: Renewables Share in Electricity Generation and New Capacity Addition, 2022

Lao PDR = Lao People's Democratic Republic, RE = renewable energy.
Source: EMBER Annual Electricity Statistics (accessed 2 July 2024).

The legacy of coal-fired power plants (CFPPs) capacity in the region is the main barrier to an accelerated clean energy transition and rapid scale-up in renewable energy capacity. Assessed countries score lowest in the average life of CFPPs compared to advanced economies and the world. This is because CFPPs in developing Asia are very young and have an average remaining life span of over 25 years (Figure 7).[16] Many countries within the region rely heavily on coal power generation as coal is a cheaper and more abundant resource. As these power plants are relatively young and provide a stable energy source, developing Asian economies will need support to secure cleaner and more reliable energy sources and address stranded asset risks.

---

16 IEA. Average Age of Existing Coal Power Plants in Selected Regions in 2020–Charts–Data and Statistics (accessed 2 July 2024).

Box 2

Uzbekistan's Solar Energy Transition

In 2019, the Government of Uzbekistan signed a 1-gigawatt solar public–private partnership (PPP) program with the Asian Development Bank (ADB) to develop and tender for multiple solar PPP projects. The first project—the Sherabad Solar PPP project—included the design, financing, construction, operation, and maintenance of a 457-megawatt solar power plant with a 52-kilometer-long transmission line near the city of Sherabad in Southern Uzbekistan.

ADB's advisory team created a government-friendly risk allocation project structure to deliver increased value for the government. In addition, the ADB team helped structure a partial credit guarantee mechanism to de-risk investment for private project developers.

Uzbekistan's renewable energy auctions achieved a tariff of $0.018 per kilowatt-hour, the lowest in Asia and the Pacific for grid-connected solar photovoltaic projects, and has become a benchmark for future solar tenders in Uzbekistan and the wider region.

Source: ADB internal communication.

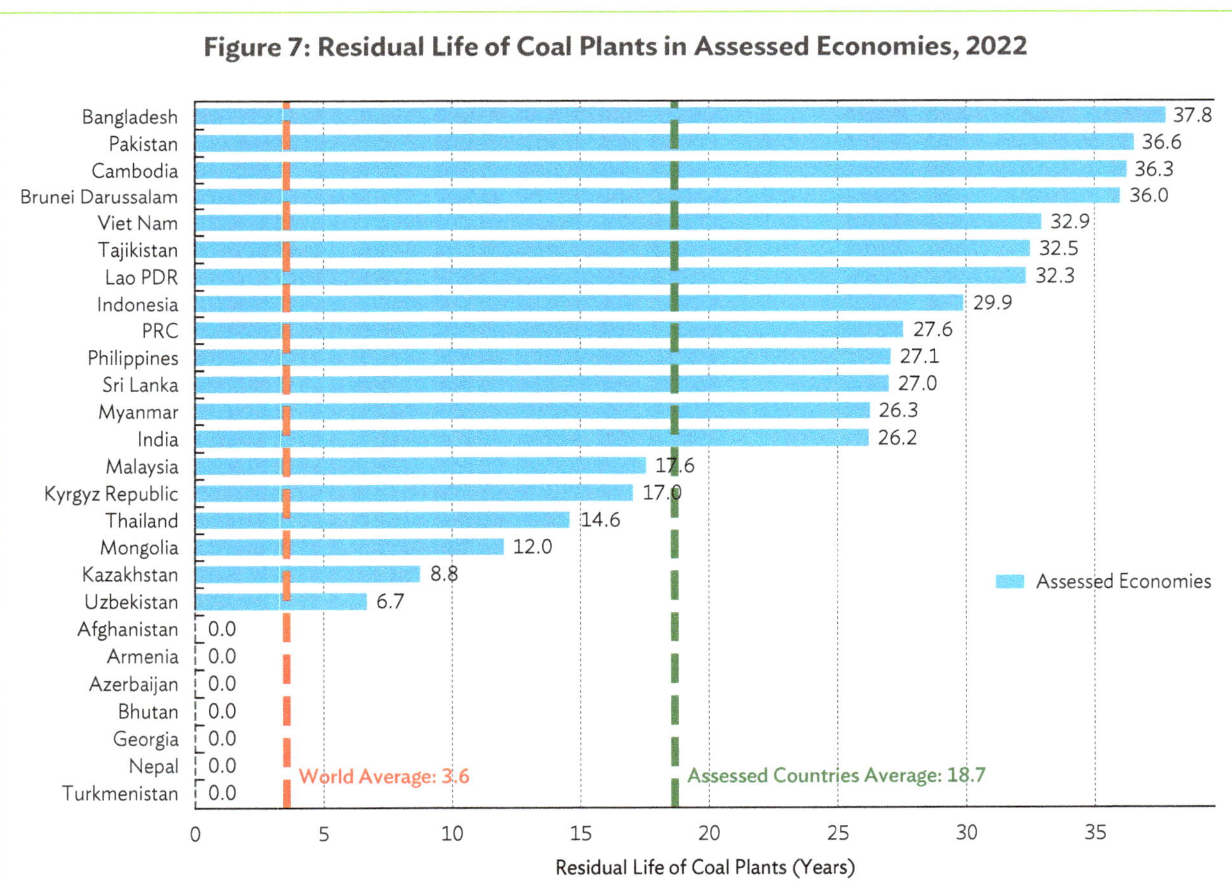

Figure 7: Residual Life of Coal Plants in Assessed Economies, 2022

PRC = People's Republic of China, Lao PDR = Lao People's Democratic Republic.

Source: Asian Development Bank calculation based on Global Energy Monitor Global Coal Plant Tracker (accessed 2 July 2024).

The early retirement of CFPPs is possible but would require appropriate carbon pricing, frameworks for transition finance, substantial amounts of concessional funding, and systems and capacity in place that facilitate a just transition for those workers and communities that would suffer adverse impacts from an earlier phaseout (Box 3). Just energy transition partnerships (JETPs) are emerging as a growing example of a regional-country platform aimed at accelerating the deployment of renewable energy and the phase-down of coal-fired electricity generation. JETPs bring together public, private, and philanthropic finance and technical assistance to accelerate and support a socially just energy transition. In developing Asia, JETPs have been announced in Indonesia and Viet Nam.

Box 3

Phaseout of Coal-Fired Power Plants

Developing Asia accounts for more than half the global greenhouse gas (GHG) emissions, most of which come from coal-fired power plants (CFPPs). By the end of 2023, legacy coal-based power plants accounted for more than 60% of electricity generation in the People's Republic of China, India, Indonesia, and the Philippines. Yet, developing Asia's coal industry is potentially declining. Of the seven Southeast Asian coal producers, only Indonesia's coal production expanded into 2024. Thailand and Viet Nam are likely to have fully depleted their remaining coal reserves by 2030, and their CFPPs—using domestic coal—could be stranded owing to a lack of alternative coal supply sources.

Challenges to the CFPP phaseout. Short-term national interests and economic priorities challenge the long-term sustainability and global ambitions of reducing GHG emissions. The average residual economic useful life of coal plants in the region exceeds 25 years. A total phaseout of CFPPs by 2050 or 2060 calls for the early retirement of several plants, which the governments of developing Asia find challenging. High termination costs of power purchase agreements—mostly with state-owned utilities— are a major factor. Termination costs of multiyear coal supply can also be substantial. An absorption of cost from coal phaseout in electricity cost would adversely affect the disposable income of low-income households. Competitively priced baseload alternatives (such as natural gas or regasified liquefied natural gas-fired power plants) do not exist in sufficient capacity.

While the clean energy transition in the region is projected to result in an overall net gain of 1.5 million jobs per year, it will also result in substantial job losses in the traditional coal sector. Moreover, the closure of CFPPs and (in some cases) coal mines feeding these CFPPs would lead to massive declines in the rural economies; small businesses would be adversely affected, and tax and nontax revenues would decline. Advancing a just transition through training and skilling schemes and strengthening social protection—including unemployment benefit schemes—can help catalyze green development by overcoming challenges faced by those who will be negatively affected.[a] Two broad streams of reducing carbon dioxide emissions from CFPPs have emerged: the carbon capture, utilization, and storage (CCUS), and the Energy Transition Mechanism (ETM).

Carbon capture, utilization, and storage. Asian governments have often cited the utilization and storage of carbon dioxide (CO_2) as a possible alternative to early retirement and essential for an orderly phaseout of CFPPs. These technologies can help reduce GHG emissions, enable CFPPs to continue their baseload generation, contribute to energy security, help retain and create jobs, and thus provide a pathway that balances economic, environmental, and social factors. Carbon capture storage and utilization have progressed slowly mainly because of the high capital costs and significant loss of output.

continued on next page

Box 3 *continued*

It is not easy to retrofit an operating CFPP with a carbon capture facility; in addition to space limitations, significant technical risks might not be acceptable to CFPP owners. Monoethanolamine production—the preferred solvent for CO_2 capture applications—has a high CO_2 footprint. The risk of a failure of CO_2 storage is also real.

A possible exception is enhanced oil recovery from depleting oil and gas fields, but incremental CO_2 emissions from the increased oil and gas recovery would need to be considered. Captured CO_2 can be used for methanol production, making chemicals (not combustion), or cement, where it is permanently sequestered.

South Asia is implementing some carbon capture utilization facilities. For example, India has completed a demonstration unit for carbon capture from CFPP flue gases and awarded an integrated CO_2-to-methanol facility for construction. In Southeast Asia, Petronas will bring a CO_2 storage site into operation in 2026, and more than 15 carbon capture storage projects are under discussion in Indonesia, mainly for enhanced oil recovery.[b]

Energy Transition Mechanism. The Asian Development Bank (ADB) launched the ETM in 2021 with the objectives of accelerating the retirement or repurposing of CFPPs and other carbon-intensive power generation (e.g., heavy fuel oil) in developing Asia, replacing them with renewable energy alternatives, and doing so in a just (equitable) manner. Core to ADB's work on ETM are (i) a just energy transition that protects the livelihoods of workers and communities affected by the transition, and (ii) carbon market development as a means to scale the accelerated retirement of thermal-power generation and meet global climate mitigation efforts.

In late 2021, ADB completed an ETM pre-feasibility study covering Indonesia, the Philippines, and Viet Nam. With the governments of Indonesia and the Philippines signifying their commitment to work with ADB to further explore how ETM can support their energy transition pathways, ADB started work on detailed feasibility assessments, covering technical, financial, legal, commercial, environmental, social, and just transition aspects, with intensive stakeholder negotiations.

At the 2023 United Nations Climate Change Conference (COP28) in December 2023, an agreement was signed between with ADB, Cirebon Electric Power, PT PLN (Indonesia's state-owned electric utility), and the Indonesia Investment Authority to shorten the power purchase agreement of the 660-megawatt Cirebon 1 CFPP by approximately 7 years.[c] ADB has also supported Indonesia in establishing the ETM Country Platform, developing Indonesia's energy transition road map and securing $500 million of highly concessional financing under the Climate Investment Funds' Accelerating Coal Transition (ACT) program. ADB has also supported operationalizing the secretariat for the Just Energy Transition Partnership (JETP), a partnership to mobilize $20 billion from the International Partners Group comprising governments and enterprises and disburse it over 3–5 years.

In the Philippines, ADB led the support for the government to access $500 million of concessional funds from the CIF ACT program, to develop independent power producer transactions and just transition programs. A detailed ETM feasibility study is about to commence in Kazakhstan. Additional countries have expressed interest to consider how ETM could contribute to energy transaction activities. ADB is working with the Monetary Authority of Singapore to scale up the ETM and develop high-integrity transition carbon credits that can serve as the foundation for accelerated thermal closures regionally and globally.

---

[a] ADB. 2023. *Asia in the Global Transition to Net Zero: Asian Development Outlook 2023 Thematic Report.*

[b] H. Mazlan. 2023. Unlocking New Opportunities Through Carbon Capture and Storage. *Petronas.* 8 February.

[c] ADB. 2023. New Agreement Aims to Retire Indonesia 660-MW Coal Plant Almost 7 Years Early. News release. 3 December.

3.3 Energy Economy Linkages

The energy economy linkages dimension assesses developing Asia's energy transition readiness to decouple energy use from economic growth based on energy trade; the nexus between energy consumption, emissions, and economic growth; and resource dependence. Figure 8 presents the results of the ETRA analysis.

Energy and emission intensities in developing Asian economies have generally fallen. The energy intensity and emission intensity median scores of assessed countries are significantly lower than those of advanced economies and world medians. This significant gap highlights the limited adoption of best-in-class and energy-efficient technologies in these regions. The lower scores indicate limited adoption of efficient energy production and consumption methods, resulting in higher energy use and emissions per unit of economic output. This inefficiency is further evidenced by their particularly low scores in the emissions intensity of exports and trade balance of low-carbon technology (LCT) products.

Adopting clean energy technologies is critical for developing Asian economies to support their net-zero transition and reduce emissions from their highly carbon-intensive industries. Industrial decarbonization will play a significant role in developing Asia's energy transition. Energy-intensive manufacturing has relocated to developing Asia and other developing countries, including the PRC and India (and is captured in the emissions intensity of exports indicator). It relies extensively on fossil fuel use—manufactured products are often meant for export—and efforts to lower emissions from such industries must be accelerated.

Given the region's dependence on trade as an engine of growth, assessed countries need to address their particularly low median score on emissions intensity of exports. New regulations in major import markets that tax the embedded carbon of export products—such as the European Union's Carbon Border Adjustment Mechanism—will impact the competitiveness of developing Asian country exports. Governments need to introduce policies and mechanisms that support the decarbonization of manufacturing and their trade. This could allow the assessed countries to leapfrog and champion decarbonization solutions. For example, India is developing solutions for maritime transport decarbonization.

Developing Asian countries are net exporters of emissions, while advanced economies export more LCT products. There has been progress in accelerating the energy transition since the Paris Agreement entered into force in November 2016, but more needs to be done. While many social, political, and other factors influence the pace of the energy transition, a key challenge is proper accounting for CO_2 and GHG emissions at the national and regional levels.

Progress on decarbonizing the fuel mix is uneven across developing Asian countries. The most striking decline has been in the PRC, driven by rapid deployment of clean energy infrastructure, shifts to manufacturing of higher value-add products, and leveraging green economy sectors as a critical pillar of economic growth.

Time-series data based on production alone does not give a complete picture of the extent to which a country has transitioned to a low-carbon path. Highly energy-intensive assessed countries can decarbonize through (i) economic transformation toward a more service-oriented economy; (ii) policy measures that encourage adoption of best-in-class energy-intensive manufacturing equipment and processes; (iii) reduction in fossil fuel subsidies; and (iv) incentives to accelerate the adoption and deployment of clean energy technologies.

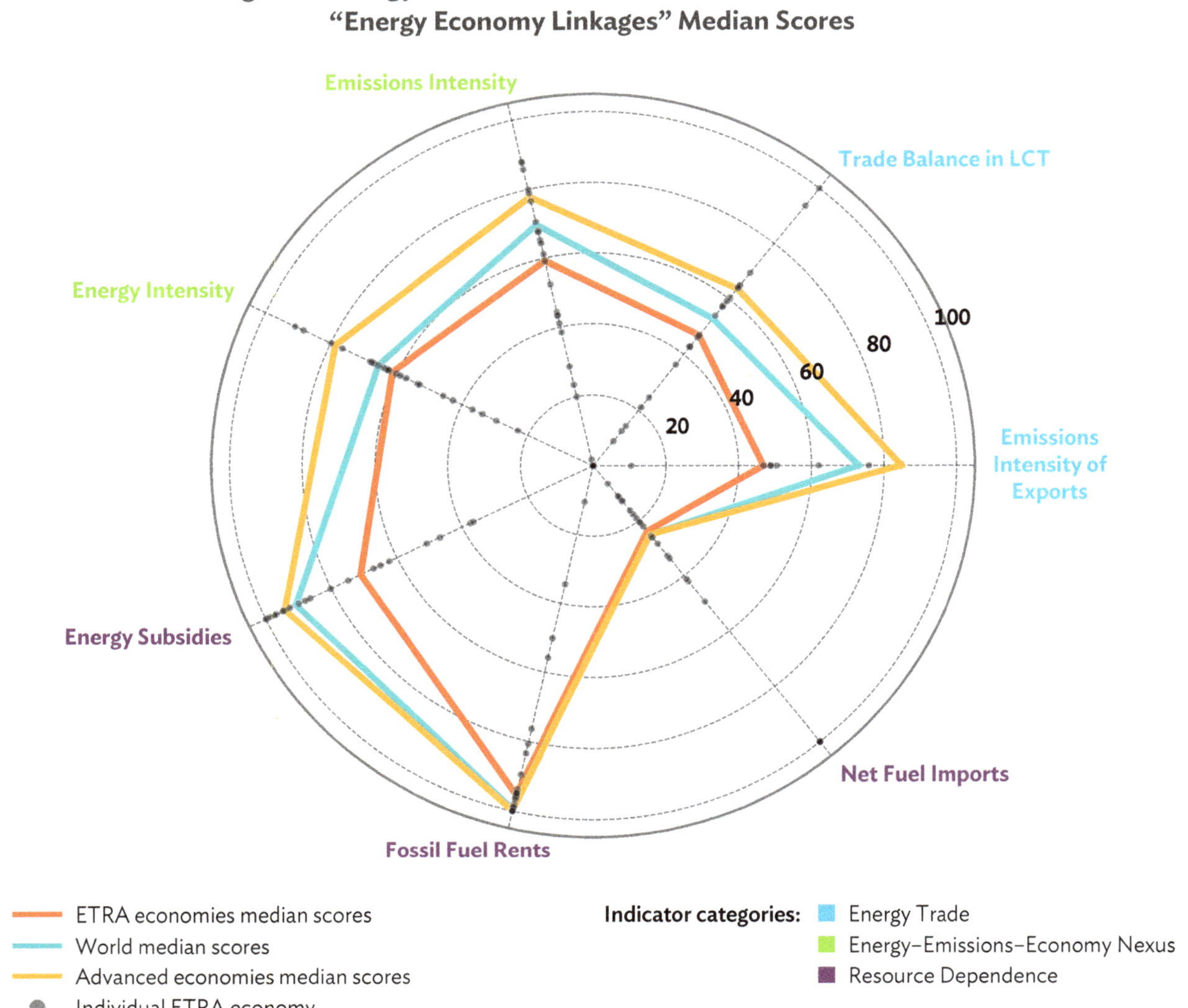

Radar Plot Highlights

- Assessed countries' lower median scores for the energy intensity and emissions intensity indicators compared to advanced economies' and the world reflect their less diversified, fossil-fuel dependent economies. The adoption of LCT and energy-efficient technologies can support developing Asia's energy transition; however, assessed countries see a lower median score for LCT products trade balance compared to advanced economies and the rest of the world, with the People's Republic of China and Malaysia as notable outliers.

- The emission intensity of exports from the assessed countries is significantly higher than the world average and that of advanced economies. This is due to the economic structure of assessed economies, partly driven by the emergence of developing Asia's energy-intensive manufacturing hub with trade linkages across the world.

- Advanced economies have been relatively more successful in eliminating or phasing out inefficient fossil fuel subsidies. With a lower-than-the-world score and nearly 25% lower than advance economies' median score, assessed countries continue to incentivize fossil fuel use, through energy subsidies, which tends to incentivize inefficient consumption of energy for fossil-fuel-importing countries.

ETRA = energy transition readiness assessment, LCT = low-carbon technology.

Note: Lower score = worse, higher score = better.

Source: Asian Development Bank.

International trade in LCT products—such as solar panels and wind turbines—has increased during the past few decades. The trade balance in LCT products (the difference between export and import values of LCT products) has been better for advanced economies (whose score is 35.8% higher than for assessed countries).[17] Among developing Asian countries, the PRC has the best LCT product trade balance and may be considered as having set a benchmark that other ETRA economies might follow. As further discussed in Chapter 3.6 (Technology and Diffusion Ecosystem), developing Asian economies must take efforts to improve their LCT trade balance by creating policy and fiscal environments that incentivize clean energy adoption (Box 4).

Box 4

India's Petroleum Subsidy Reforms—Aligning Fiscal and Climate Goals

India's phased reform of petroleum product subsidies highlights a methodical approach to balancing fiscal sustainability with energy transition readiness. In 2010, India deregulated petrol prices and moved to market-linked diesel pricing in 2014, eliminating blanket subsidies. The Pradhan Mantri Ujjwala Yojana complemented this transition by targeting vulnerable groups, providing more than 80 million liquefied petroleum gas connections to replace polluting fuels like biomass. Concurrently, excise duties on petrol and diesel were increased significantly—up to 50% of retail prices—as an implicit carbon tax to discourage consumption and fund renewable energy investments.[a] High excise duties act as a proxy carbon price, driving behavioral change and incentivizing cleaner energy adoption. This strategy supports both fiscal consolidation and India's climate commitments.[b]

India's subsidy reforms reduced fiscal spending on petroleum products from $25 billion in 2013 to $3.5 billion by 2023, paving the way for investments in renewables and infrastructure.

---

[a] World Economic Forum. (Gradual) Removal of Fossil Fuel Subsidies.
[b] Asian Development Bank. 2023. *Asia-Pacific Climate Report 2024: Catalyzing Finance and Policy Solutions.*

While having made good progress in reducing fossil fuel subsidies until 2020, with rising fuel prices in the international markets, developing Asian countries have since increased fossil fuel subsidy levels. Assessed countries score nearly 25% lower than advanced economies' median score. Data indicate that fuel subsidies tend to increase when international prices are high and reduce when global prices are low so that governments can keep energy prices low for consumers. Maintaining fossil fuel subsidies can lead to significant fiscal constraints, encourage dependence on fossil fuels, and result in the inefficient allocation of resources necessary for long-term energy transition planning (Figure 9). While progress is slow, assessed countries are beginning to adopt carbon pricing regimes (Box 5). To enhance the effectiveness of their climate action, countries will need to rationalize their subsidy and tax regimes.

---

[17] IMF. 2024. *Trade in Low Carbon Technology Products.* Climate Change Dashboard. 29 November (update).

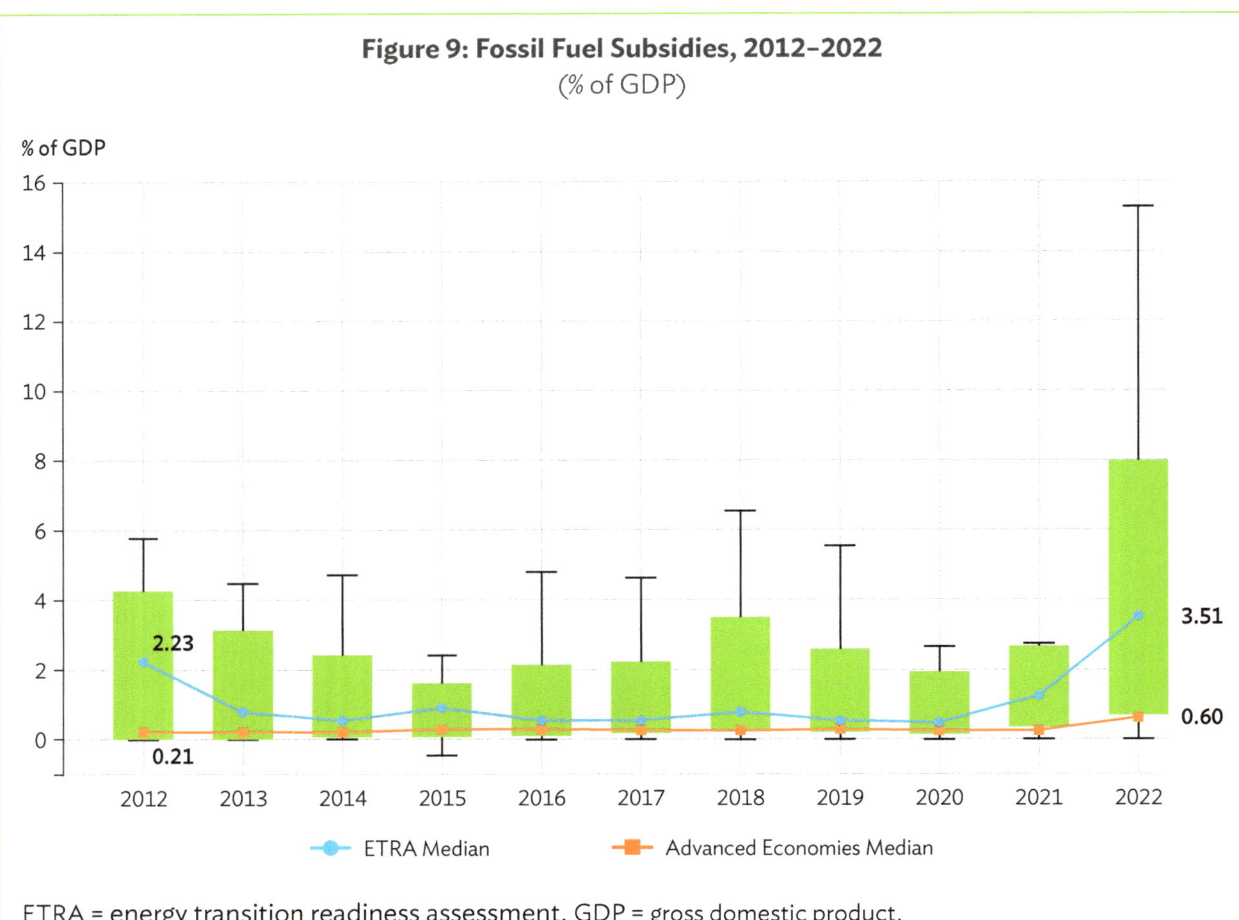

ETRA = energy transition readiness assessment, GDP = gross domestic product.
Source: Fossil Fuel Subsidy Tracker (accessed 2 July 2024).

Box 5

Carbon Pricing in Asia and the Pacific

The momentum to utilize carbon pricing instruments (CPIs) in Asia and the Pacific continued to grow in 2023/2024. This momentum included considering, adopting, and/or further developing a range of CPIs: carbon taxes and emissions trading systems (ETSs), domestic carbon crediting mechanisms, and international carbon markets under Article 6 of the Paris Agreement and the voluntary carbon market (VCM).

Carbon Taxes and Emissions Trading Systems

As of 2024, there were nine direct CPIs in operation in Asia and the Pacific: a carbon tax in Taipei,China; Japan (at the subnational level); and Singapore; and an ETS in Australia, the People's Republic of China (PRC), Indonesia, Kazakhstan, the Republic of Korea, and New Zealand. Launched in 2023, Indonesia's ETS is initially covering coal-fired power plants. ETSs also exist at the subnational level: Japan has the Tokyo and Saitama ETSs at the subnational level, while the PRC is integrating eight pilot ETSs across two cities and six provinces into the national ETS.

continued on next page

Box 5 *continued*

CPIs are being considered or are under development in Brunei Darussalam, Japan (at the national level), Pakistan, the Philippines, Thailand, and Viet Nam. In early 2024, Japan approved the basic plan for the Ministry of Economy, Trade and Industry's *GX: Green Transformation Policy*, under which Japan will finalize guidelines for its national ETS by 2026/2027 and introduce carbon levies on fossil fuel importers such as refiners, trading houses, and electricity utilities by 2028/2029.[a] In 2023, the Malaysian state of Sarawak approved legislation to introduce a carbon tax to reduce emissions and promote carbon capture and storage. Thailand and Viet Nam are preparing legislation for domestic ETSs.

Governments allow regulated entities to use carbon credits toward their greenhouse gas (GHG) reduction obligations to increase flexibility, lower compliance costs, and extend the carbon price signal to uncovered sectors. Singapore began allowing businesses liable to pay the carbon tax to use international carbon credits (under Article 6) that meet defined environmental integrity criteria to offset up to 5% of their taxable emissions in January 2024.[b] The ETS in the Republic of Korea allows for the limited use of carbon credits from specified crediting mechanisms.

Domestic Crediting Markets

Several countries have or are developing domestic carbon crediting mechanisms. In 2024, the PRC officially relaunched its China Certified Emission Reduction program, which allowed carbon credits to be generated from projects in offshore wind, solar thermal power, mangrove development, and forestry. Voluntary buyers may purchase the generated credits and can cover up to 5% of verified emissions of entities covered by the PRC's national emission trading system.[c] A similar approach is also being used in Indonesia, which allows domestic carbon credits to be purchased voluntarily or used by entities domestically under certain sector conditions to meet GHG emission targets or related obligations.

In 2024, Thailand introduced updates to its scheme to ensure a coherent framework is in place for domestic crediting activities, which includes clear identification of eligible project domains and delineation of different credit types depending on the standards under which credits are developed. India is implementing the Indian Carbon Market to facilitate the achievement of its enhanced nationally determined contribution targets. A GHG emission intensity-based compliance mechanism for regulated entities and a carbon credit trading scheme (CCTS) underpins the Indian Carbon Market.[d]

Entities not covered by compliance mechanisms can also implement projects to generate carbon credits for the CCTS or under the VCM.

International Carbon Markets

Under Article 6 of the Paris Agreement, more countries are preparing for engagement by developing regulatory frameworks for international carbon markets. A driver in this development is the emergence of bilateral transfer agreements under Article 6.2. As of 2 August 2024, there were 86 bilateral agreements between 11 different buyers and 44 host countries, with 140 pilot projects, 119 belonging to Japan's Joint Crediting Mechanism.[e]

An internal preliminary review by the Asian Development Bank of readiness for Article 6 identified that its developing member countries exhibit different levels of progress. Out of the 16 countries assessed, only one country had all the essential elements in place or under development to effectively operationalize Article 6 opportunities; eight countries had put in place legislative arrangements, or were doing so but lacked a coherent strategy or had significant capacity gaps; and seven countries were either in early or nascent stages of putting in place the necessary elements to start operationalizing opportunities.

continued on next page

Box 5 *continued*

Trading platforms are being set up in Malaysia, Indonesia, Singapore, and Viet Nam, but many countries have not yet implemented the infrastructure to comply with Article 6.

As of August 2024, preliminary estimates indicate that more than 70% of the cumulative global demand for internationally transferred mitigation outcomes for meeting nationally determined contribution targets up to 2030 will originate from Asia and the Pacific.[f] Starting in 2027, if all major Asian economies join the Carbon Offsetting and Reduction Scheme for International Aviation, airlines registered in Asia and the Pacific are expected to have the highest aggregate offset responsibility of any region.

Asia and the Pacific continues to be the largest suppliers of carbon credits to the VCM in terms of volume and value. In 2023, about 23.1 metric tons of carbon-dioxide equivalent ($mtCO_2e$) credits originated in the Asia and Pacific region for $5.55 per tons of carbon-dioxide equivalent (tCO_2e), the lowest average price of any region.[g] VCM transactions from projects in Asia experienced sharp drops in volume and price, impacted by structural shifts taking place in the broader global carbon market. The voluntary market in Asia and the Pacific is generally served by independent carbon programs that operate their registry systems, with transactions usually handled through established exchanges or over-the-counter trades. The emergence of national carbon crediting mechanisms and increasing synergies between compliance and voluntary markets create an increasing need for coherence between voluntary and compliance-based systems.

---

[a] Y. Obayashi and K. Golubkova. 2023. Explainer: Japan's Carbon Pricing Scheme Being Launched in April. *Reuters*. 31 March.

[b] Government of Singapore, Ministry of Sustainability and Environment. 2023. Singapore Sets Out Eligibility Criteria for International Carbon Credits Under the Carbon Tax Regime.

[c] Yi Wu. 2024. Understanding the Relaunched China Certified Emission Reduction (CCER) Program: Potential Opportunities for Foreign Companies. *China Briefing*. 3 May.

[d] *International Carbon Action Partnership*. 2024. India Adopts Regulations for Planned Compliance Carbon Market. 2 September.

[e] Carbon Market Institute. 2024. *International Carbon Market Update: States and Trends in the Asia Pacific*.

[f] Asia-Pacific Economic Cooperation. 2022. *APEC Energy Demand and Supply Outlook*. 8th ed. Vol. 1. Asia Pacific Energy Research Centre.

[g] *Bain & Company*. 2023. Voluntary Carbon Markets in 2023: A Bumpy Road Behind, Crossroads Ahead.

High fossil-fuel rent demonstrates a high transition risk for developing Asian economies, advanced economies, and the rest of the world. As a share of GDP, advanced economies earn a lower economic rent on fossil fuels. The advanced economies have relatively higher net fuel imports, with a median score of 5% higher than that of the assessed countries. Within the assessed countries, Azerbaijan earns the highest fossil fuel rent and relies least on fuel imports. Despite earning high rent from fossil-fuel exports, developing Asian economies face high transition risk because of the potential for stranded assets. Those economies may adjust their fiscal regimes, accelerate the energy transition, and avoid distressed and stranded assets. For countries reliant on fossil fuel revenues for their national budgets, shifting away from fossil fuels could lead to significant fiscal deficits and economic instability. Governments must weigh the costs and benefits of delaying the energy transition and avoiding the phaseout of fossil fuel assets. Early adoption and diversification of the economy toward a shift in renewable energy deployment and clean energy technology manufacturing can create alternative and sustainable jobs within a green economy.

3.4 Infrastructure System

The transformation of energy infrastructure is central to the energy transition, requiring developing Asian countries to shift from centralized toward more resilient, decentralized, and flexible energy systems. The infrastructure system dimension comprises three subdimensions: (i) climate resilience and coping capacity, (ii) energy infrastructure quality, and (iii) physical and digital infrastructure. These subdimensions are categorized into seven indicators (Figure 10).

Developing Asian economies must manage energy infrastructure vulnerabilities to facilitate the energy transition. Analysis shows that the coping capacity of energy infrastructure to handle climate shocks in developing Asian countries is about two-thirds that of advanced economies, and the energy sector is also vulnerable to climate change-induced water stress exacerbated by water unpredictability and possibly scarcity at a time when the need for reliable water supplies is rising. The assessed economies' median score in the "water stress" indicator shows that overall, developing Asian countries are more water-stressed than advanced economies (Box 6). Cambodia is the least water-stressed because more than 80% of its land area is in the Mekong River basin; it has abundant surface water and aquifer resources and a high level of rainfall. Given its economic strength, Brunei Darussalam is well-positioned to cope with climate-related weather events. The Brunei Darussalam government has set ambitious targets for reforestation and afforestation programs to help address the risks of flooding and landslides. In addition to climate change-induced vulnerabilities, energy infrastructure is vulnerable to cyberattacks. Each type of threat presents unique challenges that must be addressed to keep these vulnerabilities within manageable limits.[18]

Box 6

Solar Irrigation in Bangladesh—Advancing the Energy–Water Nexus

Bangladesh's solar irrigation initiatives illustrate a positive water–energy nexus, demonstrating how renewable energy can enhance freshwater sustainability in developing Asia. Bangladesh's Solar Irrigation Road Map (2023–2031), supported by the Asian Development Bank, integrates solar-powered irrigation pumps (SIPs) to replace diesel pumps. Targeting 45,000 SIPs will displace 300,000 tons of diesel annually while reducing carbon emissions by 900,000 tonnes. SIPs prioritize surface water irrigation, reducing pressure on depleting groundwater resources and ensuring long-term agricultural resilience. Further, it was estimated that 1.3 million farmers will benefit from a 50% reduction in irrigation costs, saving $377 million annually. Around 480 gigawatt-hours of excess electricity generation can feed into the national grid, bolstering rural electrification.

Source: Asian Development Bank. 2023. *Road Map to Scale Up Solar Irrigation Pumps in Bangladesh (2023–2031)*.

---

[18]   Senstar. Energy Sector Vulnerabilities.

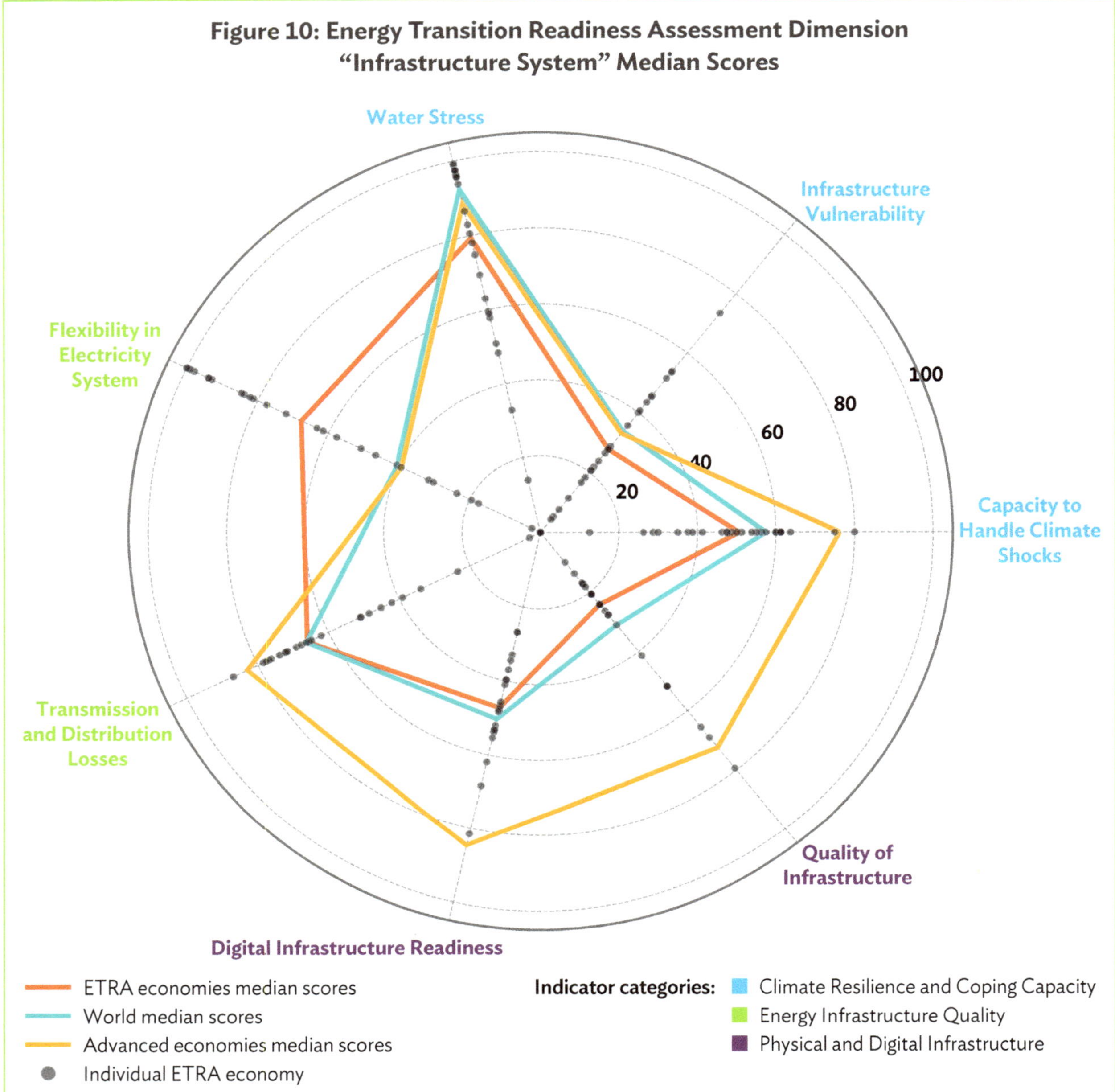

Figure 10: Energy Transition Readiness Assessment Dimension "Infrastructure System" Median Scores

Radar Plot Highlights

- Developing Asian countries are more vulnerable to disasters and climate-related events than advanced economies, and their coping capacity is significantly less, with a capacity to handle climate shock score about two-thirds that of advanced economies. This is often due to underinvestment and inadequate maintenance of their energy infrastructure, characterized by high transmission and distribution losses and low availability.

- Developing Asian countries water stress score is 10% below advanced economies' median, indicating higher vulnerability and limited capacity to cope with climate change-induced water scarcity.

- Developing Asian economies must adapt and invest in digital infrastructure and literacy to enhance digital readiness for the energy transition.

ETRA = energy transition readiness assessment.

Note: Lower score = worse, higher score = better.

Source: Asian Development Bank.

Strong transmission and distribution (T&D) infrastructure—with low commercial and technical network losses—reinforces the quality of power delivery systems and demonstrates the readiness to electrify developing Asian countries with renewables, but further progress is needed. Electricity loss reduction means that less electricity is required to meet demand. Developing Asian countries must reduce losses in transformers, lines (conductors), and other equipment and improve the power factor. The "T&D losses" indicator shows that the assessed countries' network loss levels are substantially higher than for advanced economies because of overloaded lines and undersized transformers resulting from inadequate planning, unplanned expansion, and poor maintenance practices (Figure 10). Assessed countries made some progress in reducing T&D losses from 2000 to 2021 (Figure 11), and access to grid electricity also increased during the same period (Figure 12).

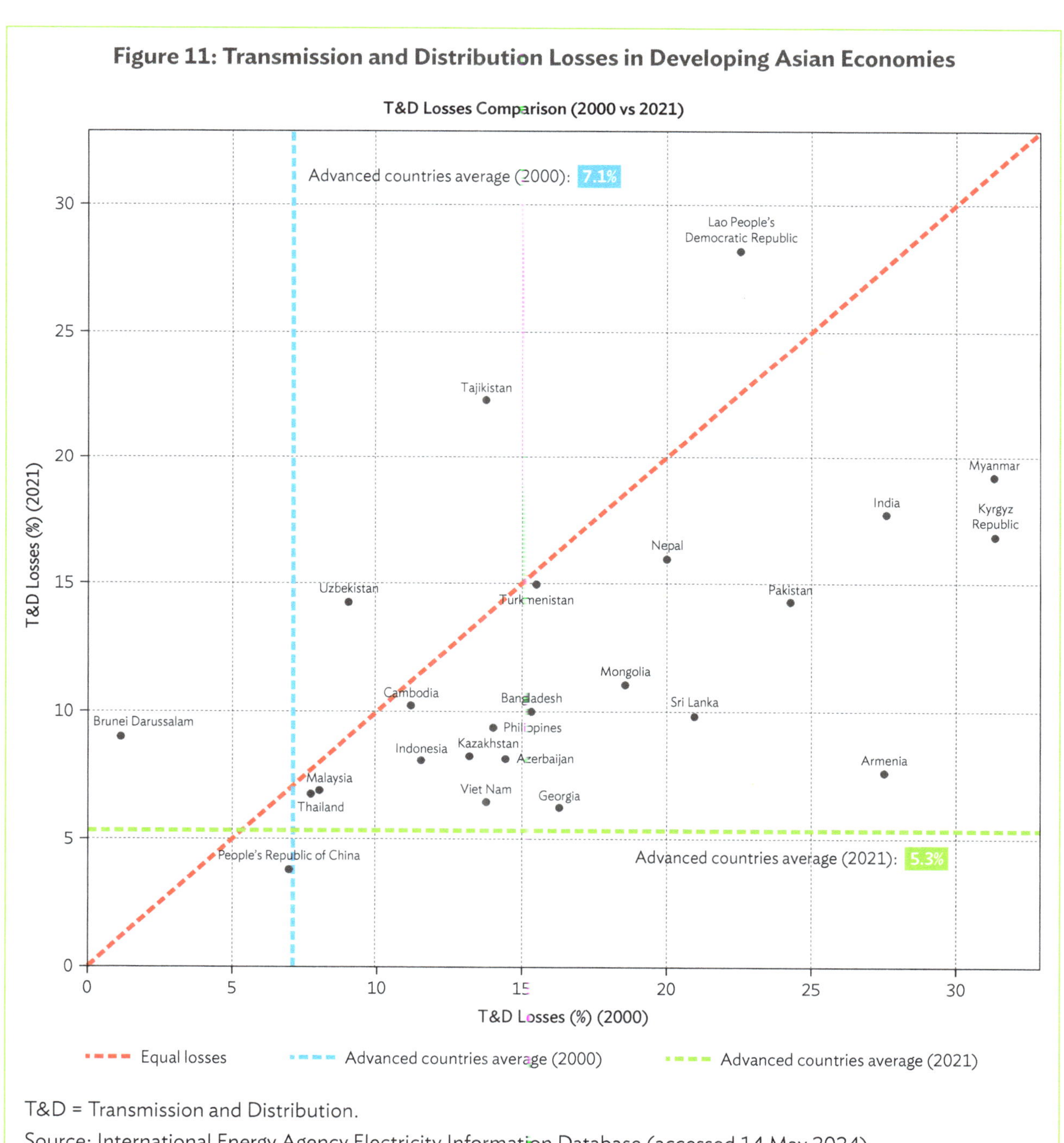

Figure 11: Transmission and Distribution Losses in Developing Asian Economies

T&D = Transmission and Distribution.

Source: International Energy Agency Electricity Information Database (accessed 14 May 2024).

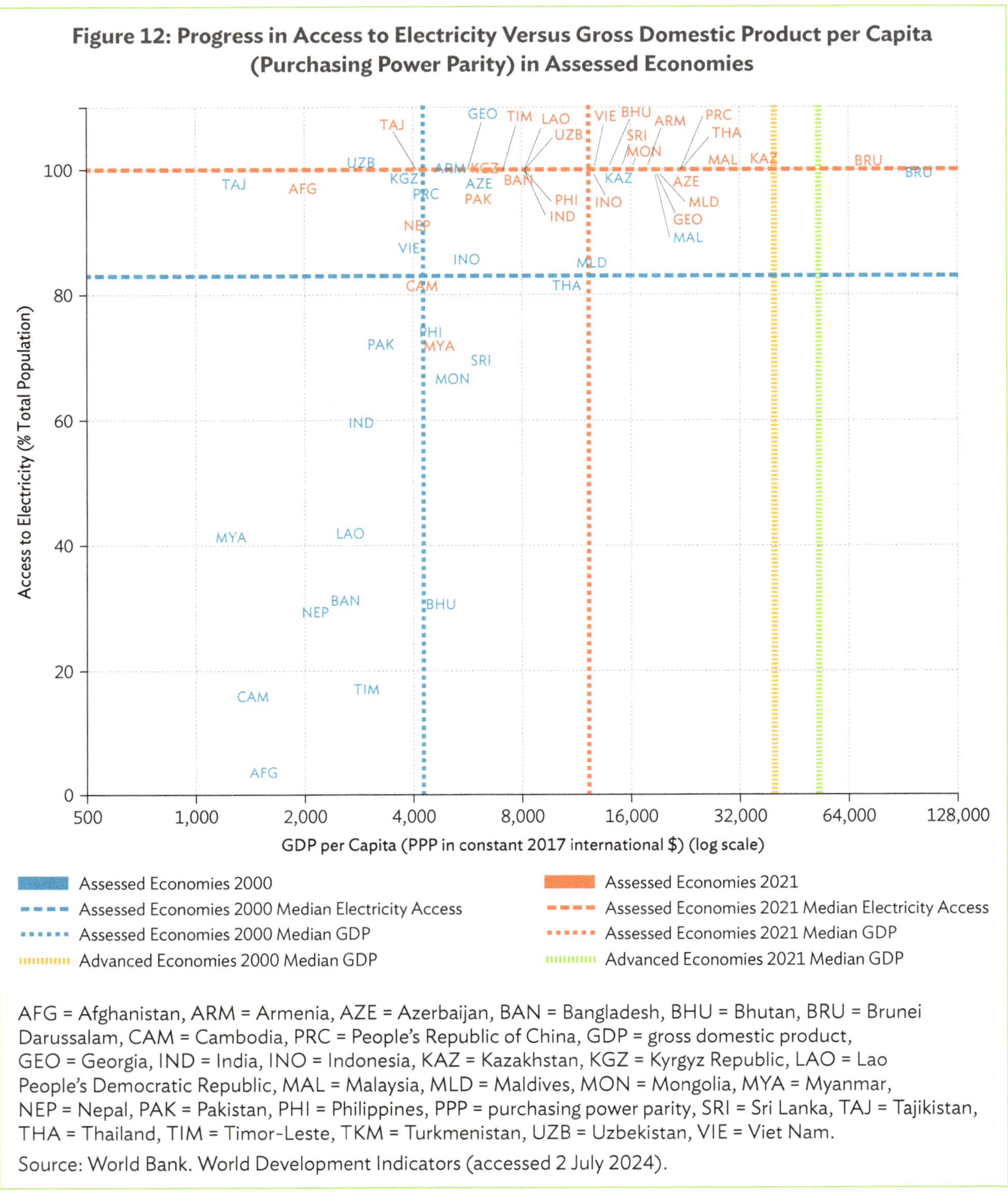

Figure 12: Progress in Access to Electricity Versus Gross Domestic Product per Capita (Purchasing Power Parity) in Assessed Economies

AFG = Afghanistan, ARM = Armenia, AZE = Azerbaijan, BAN = Bangladesh, BHU = Bhutan, BRU = Brunei Darussalam, CAM = Cambodia, PRC = People's Republic of China, GDP = gross domestic product, GEO = Georgia, IND = India, INO = Indonesia, KAZ = Kazakhstan, KGZ = Kyrgyz Republic, LAO = Lao People's Democratic Republic, MAL = Malaysia, MLD = Maldives, MON = Mongolia, MYA = Myanmar, NEP = Nepal, PAK = Pakistan, PHI = Philippines, PPP = purchasing power parity, SRI = Sri Lanka, TAJ = Tajikistan, THA = Thailand, TIM = Timor-Leste, TKM = Turkmenistan, UZB = Uzbekistan, VIE = Viet Nam.

Source: World Bank. World Development Indicators (accessed 2 July 2024).

As the energy transition progresses with increased penetration of distributed and variable renewable energy sources, grid services and power distribution management become more challenging. Utilities in developing Asian countries are trying to keep pace with the evolving needs, but their energy infrastructure quality is lagging, with a score of about a third of advanced economies, mainly because of inadequate maintenance expenditures and practices.

Good quality power will facilitate the energy transition in developing Asian countries. Low-voltage fluctuations, harmonics, and transients help reduce downtime of power-using equipment, reduce fault conditions, and ensure sustainable, safe, and efficient utilization of electrical systems. These factors will increase distributed and variable renewable energy penetration in power grids. The energy transition objectives will also be furthered when it becomes possible to take advantage of price fluctuations on power exchanges, i.e., by shifting consumption to when tariffs are low and/or shifting production to when tariffs are high.

Developing Asia's energy transition readiness can be accelerated by strengthening power system flexibility and fostering the adoption of digital solutions. The score on the "flexibility of electricity system" indicator of developing Asian economies is better than that of advanced economies and reflects their emphasis in recent years on developing hydropower resources and setting up gas-fired capacity. With a large share of gas-fired capacity, the power system in Turkmenistan is the most flexible among the developing Asian countries.[19] Owing to the anticipated increases in decentralized electricity production, the need for flexibility services is also expected to become increasingly important in the medium term.[20]

In an increasingly digitally driven world that enables system flexibility and power quality, digital technologies will become increasingly critical to power system operations.[21] Developing Asian countries must accelerate the energy transition by investing in digital infrastructure. This infrastructure comprises the internet backbone, mobile telecom, digital communication, and data centers (among other aspects).[22] Digital technologies will help increase responsiveness to price signals, grid frequency movements, and signals from grid operators to adjust electricity consumption or production to balance electricity demand with supply and to ensure grid stability at all times (Box 7).

Box 7

Grid Transformation Is Central to Energy Transition

Economic development, income growth, and government commitment have resulted in significant progress in increasing electricity access in developing Asia over the past decades. In 2021, the average access to electricity rate exceeded 97%, up from about 70% in 2000.

Grid losses have decreased from an average of about 14% in 2000 to 12% in 2022. Yet, developing Asia's grid infrastructure is not sufficiently prepared for accelerating the energy transition. Supply quality and high overall technical losses are manifestations of insufficient grid preparedness. An updated, expanded, and digitalized grid infrastructure is essential to seamlessly integrate diverse and distributed low-carbon electricity generators and enable increased electrification and final energy demand profile optimization.

continued on next page

---

[19]   Like in other regions, natural gas is also considered a transition fuel in ETRA countries.

[20]   Next Kraftwerke. What is Flexibility in the Electricity Sector?

[21]   ADB. 2012. *Climate Risks and Adaptation in the Electric Power Sector*; and Shenzhen Clou. 2023. What is Power Quality and Why is it Important? *Smart Energy International*. 18 April.

[22]   L. Ogbevoen. n.d. What is Digital Readiness? Oden Technologies.

Box 7 *continued*

Electricity grid investments must be anchored to the long-term goals of the Paris Agreement and net-zero emissions. As per the International Energy Agency's net-zero emissions scenario, global electricity grids must expand annually by about 2 million kilometers by 2030.[a] Estimated grid investment of $700 billion–$800 billion per annum is needed until 2030 and beyond.[b] The most conservative estimate for the region's grid investment requirements is $2 trillion over a decade, plus another $400 billion at least in grid-connected energy storage (with 75% going to pumped hydro and 25% to batteries).[c] Developing Asia is expected to be at the forefront of grid expansion investments. The People's Republic of China is estimated to account for about 30% of the estimated global grid investment.[d]

Developing member countries (DMCs) of the Asian Development Bank (ADB) have several options to improve their grid infrastructure capabilities while striving to improve supply quality. These include (i) enhancing grid planning practices, (ii) accelerating regional and in-country interconnectivity, (iii) fostering the adoption of digital solutions, and (iv) developing effective financing mechanisms.

Integrating clean energy transition goals in grid planning. Traditional grid investment planning studies typically consider the goals of reliability, security, affordability, and cost-effectiveness, but climate goals also make it necessary to coordinate grid planning with other sectors and applications (such as electric vehicles and charging infrastructure). This goes beyond the scope of traditional grid planning studies. Some countries have conducted additional studies that guide electricity sector development in line with longer-term climate targets. For example, Perusahaan Listrik Negara—the Indonesian power utility— prepared a generation and grid development plan to become carbon neutral by 2060. The Electricity Generating Authority of Thailand—which updates (or prepares afresh) the power development plan once every 1 to 2 years—was (as of November 2023) updating the plan in line with the country's net-zero emissions pledge.

Several countries have sought to identify regions with high potential for variable renewable energy (VRE) that can be linked to major urban and industrial centers. An example is the green corridor development in India (comprising 765-kilovolt and 400-kilovolt lines and substations) that increased electricity access and private investments in VRE.

Grid integration to enhance grid flexibility. Interconnection of power systems across countries enhances grid stability and allows areas with a high renewable resource base to transfer electricity to areas with high demand and/or less generation.[e] Such electricity transfers lead to market optimization. High-voltage direct current technology can be deployed where it is impossible to maintain the same grid frequency in interconnected regions.

In August 2023, the energy ministers of the Association of Southeast Asian Nations (ASEAN) called for expanding the ASEAN Power Grid (APG) and envisioned the APG as a "digitized transmission network where renewable energy supply can be channeled to support the regional demand." The APG will enable power trade within the region, and help meet ASEAN's key strategic priorities of strengthening energy security and shifting to renewables.[f] ADB DMCs of the Central Asia Regional Economic Cooperation and South Asia Subregional Economic Cooperation regions are also working toward increasing connectivity and power transfers among their member countries. Countries with widespread geographies also interconnect power systems serving different parts of the country to realize similar benefits.

Increased grid flexibility through digitalization. Electricity grids in ADB DMCs need to modernize and upgrade to address renewables integration objectives, reduce losses and emissions, and be smart, which can be facilitated by digital technologies that automate power system control and increase

continued on next page

Box 7 *continued*

monitoring and situation awareness. Some ADB DMCs have invested in digital technologies to enhance flexibility and improve grid performance. For example, the Kyrgyz Republic deployed state-of-the-art fast-charging battery technology in more than 100 buses. These batteries were to recharge only during off-peak hours so that additional investment in the distribution network would be avoided or delayed.

Effective and innovative financing mechanisms. A range of mechanisms that unlock finance from multiple sources is required to meet grid investment needs. ADB has helped catalyze grid investments using market-based instruments and public–private participation arrangements. Private participation is particularly beneficial if contingent liabilities are well managed. ADB has provided long-tenure loans to a utility in Georgia and two city-level utilities in India (one in local currency) when financing from commercial sources on such terms was not readily available. As transmission is normally state-owned, an infrastructure investment trust facility in India allows private parties to invest in specific transmission assets and reduce transmission enterprise reliance on budgetary support. The first transmission project under the facility was operationalized in April 2024, and bids for 10 more private transmission projects are being sought.

Recommendations. The pace of grid development in ADB DMCs must be accelerated to keep up with rising demand and ensure a secure, reliable, cost-effective energy transition. Digital technologies must be deployed to enable VRE integration, enhance grid flexibility, and balance generation and load. A suitable policy, regulatory, and investment environment must be in place. This will require good and insightful leadership that steers governments, utilities, and other stakeholders toward (i) coordinating grid planning across power sector segments, as well as other economic sectors within the country and in neighboring regions; and (ii) addressing barriers to grid development, which calls for unlocking investment and financial innovation, securing supply chains for equipment and materials, leveraging digitalization with artificial intelligence and machine learning tools, and developing institutional capacities to manage the transition and beyond.

[a] The International Energy Agency does not provide a separate estimate for investment in transmission and distribution grids but includes it in the estimate for clean energy investment. This clean energy investment is expected to increase from about $1.8 trillion in 2023 to about $4.5 trillion per year by the early 2030s. This includes concessional funding, which rises by the early 2030s to about $80 billion–$100 billion per year for Asia and the Pacific and other emerging markets and developing economies. International Energy Agency. 2023. Net Zero Roadmap: A Global Pathway to Keep the 1.5C Goal in Reach—2023 Update.

[b] International Renewable Energy Agency. 2023. Tripling Renewable Power and Doubling Energy Efficiency by 2030: Crucial Steps Towards 1.5°C. Bloomberg estimates that an investment level of $4.84 trillion from 2023 to 2030 in electricity grids and flexibility is consistent with limiting global temperature rise to 1.5°C. At least $21.4 trillion would be required across the globe by 2050 to support a net-zero trajectory, which would more than triple the annual investment levels from $274 billion in 2022 to $871 billion in the decade preceding 2050 (BloombergNEF. 2023. New Energy Outlook 2022: Global Net Zero Will Require $21 Trillion Investment in Power Grids. Blog. 2 March). Rystad Energy estimates an investment need of $3.1 trillion from 2024 to 2030 (Asian Power. 2024. Asia to Lead Grid Expansion Investments. 23 February).

[c] G. Thompson. 2023. Asia's Renewables Need Massive Grid Investment. *Wood Mackenzie*. 25 May.

[d] *Asian Power*. 2024. Asia to Lead Grid Expansion Investments. 23 February.

[e] Where it is not possible to maintain the same grid frequency in the interconnected regions, the high-voltage direct current technology is deployed.

[f] *ADB Southeast Asia Development Solutions*. 2023. Energy Ministers Push for Expansion of ASEAN Power Grid and Trans-ASEAN Gas Pipeline. 7 November.

The PRC is the most advanced among the assessed countries regarding digital readiness. Overall, the median score for the digital infrastructure readiness indicator for developing Asian countries is marginally better than the world median score but substantially lower than that of advanced economies.

In addition to grid transformation, the readiness for adopting digital technologies in terms of people, processes, and technology—coupled with robust and good quality infrastructure—can enable the energy transition in developing Asian countries, such as Malaysia (Box 8). A comprehensive digital readiness assessment can be a good starting point for gauging power infrastructure quality and how much power systems can benefit from emerging digital communication, digital metering, grid management and tracking, artificial intelligence and machine learning tools, and other digital technologies. A thorough assessment would consider the organizational culture, including how people operate, maintain, and construct power networks. This could also cover the organization's processes and systems, including technical operations, business processes, and technology, which is key to automating processes, making adoption scalable, and improving compliance. An assessment of supply chains for equipment and materials will also be an essential part of this exercise.

Box 8

Digital Transformation—Malaysia

Malaysia is actively integrating digital technologies to enhance the efficiency, reliability, and sustainability of its energy sector. Malaysia's smart meter rollout demonstrates its commitment to modernizing energy infrastructure and positioning the country as a leader in Southeast Asia's energy transition.[a]

In 2019, Tenaga Nasional Berhad (TNB), Malaysia's national power utility, initiated a plan to install around 9.1 million smart meters across Peninsular Malaysia by 2026, supporting the government's vision for a digitalized energy infrastructure.[b]

As of August 2024, TNB installed around 2.3 million smart meters nationwide. Key areas include Melaka (380,000), Selangor (1.8 million), Kuala Lumpur (1.3 million), and Penang (380,000). This reflects the growing acceptance of smart technology among Malaysian consumers to manage energy use, reduce consumption, and save costs. In addition, the real-time data also enhances grid management and facilitates renewable energy integration.[c]

[a] *Bernama.* 2024. Power Smart Meters in 9.1 Million Households by 2026. 27 September.
[b] *Energy Watch.* 2022. The Go-to-Guide for Malaysia's Smart Meter. 23 December.
[c] *Bernama.* 2024. Smart Meter Technology Transforms National Energy Management. 3 August.

3.5 Social System

Universal accessibility to affordable and reliable energy has been a critical enabler of progress on key development objectives such as improving health care, food security, education, and creating employment opportunities. As clean energy becomes increasingly cost-competitive and more abundantly available, it has the potential to accelerate social development. Nevertheless, the imperative to continue the pursuit of developmental goals can, at times, compete with the necessity to decarbonize at speed and scale. These trade-offs are particularly evident in developing Asia, which accounts for 52% of the world's population, with millions of people still living below the poverty line and widening income inequality. It is essential to ensure that the restructuring of the energy system does not create negative social externalities and leaves no one behind.

The ETRA social system dimension considers a country's capabilities to ensure access to affordable energy and energy services and institutional mechanisms to identify and address the distributional implications of the energy transition on vulnerable stakeholder groups. Figure 13 shows a snapshot of the analysis for the social system dimension, which includes three subdimensions: energy affordability, energy and transportation access, and environmental health and climate vulnerability. The subdimensions are further categorized into 11 indicators.

A just and equitable transition for developing Asia must be enabled with supporting policy measures, institutional mechanisms, and necessary multistakeholder and multisector mandates. With the adoption of the "Just Transition Work Programme" at COP28 in the United Arab Emirates, prioritizing just and equitable transitions has emerged as the key to achieving net-zero goals. Similarly, the announcements of plurilateral mechanisms, such as the just energy transition partnerships (JETPs) in Indonesia and Viet Nam, have brought just and equitable transitions to the mainstream. As of 2023, only four of the assessed countries—Indonesia, Pakistan, the Philippines, and Viet Nam—mentioned "just transition" in their NDCs.[23] The principles of justice and equity are central to the energy transition and need to be acknowledged accordingly in national climate action plans. Delivering a just and equitable transition for developing Asian countries relies on understanding the extent and depth of potential social impacts and the intentional design of energy policies that do not exacerbate structural inequalities. This entails identifying types and sizes of potentially vulnerable groups, channels of impacts and severities thereof, and stakeholder engagement mechanisms at local, sector, and national levels.

Developing Asia has achieved major milestones in household access to electricity.[24] This expansion has extended to rural communities, with the share of the rural population with access to electricity reaching close to or 100% in almost all assessed countries.[25] This significantly impacted socioeconomic development, given the role electricity plays in delivering services such as health care, education, and nutrition. The progress on electrification has been enabled primarily by public expenditure on augmenting power generation capacity and expanding grid connectivity to underserved areas. To realize the social dividends of electrification, further efforts are necessary to provide capabilities toward increasing levels and diversity of end consumption.

---

[23]   T. Fransen et al. 2023. 9 Things to Know About National Climate Plans (NDCs). *World Resources Institute*. 7 December.

[24]   IEA. 2024. SDG 7: Data and Projections—Access to Electricity.

[25]   However, a significant number of people, while being connected to the grid, suffer from low-quality power.

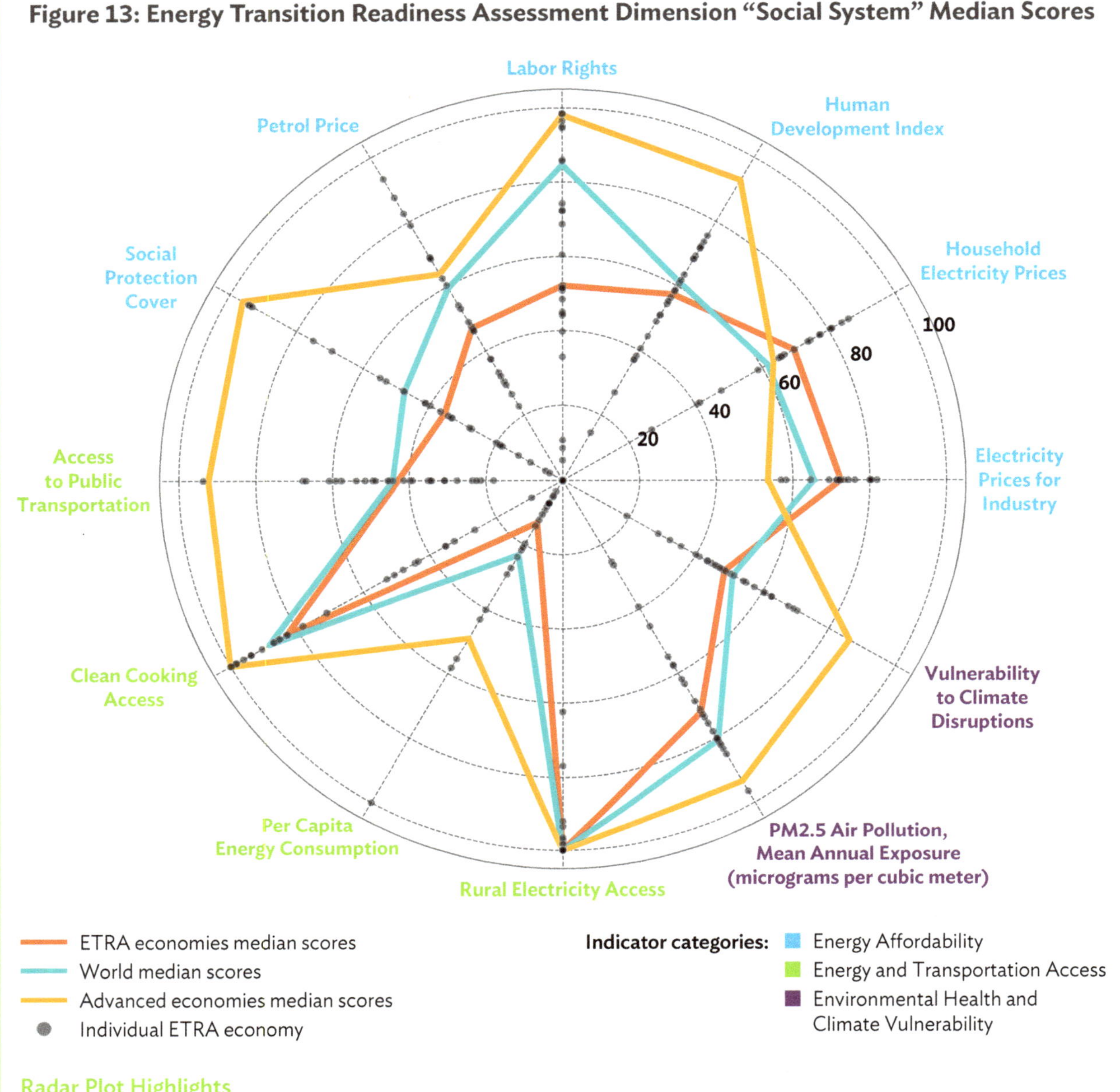

Figure 13: Energy Transition Readiness Assessment Dimension "Social System" Median Scores

Radar Plot Highlights

- Developing Asia has excelled in providing household access to electricity, achieving nearly 100% rural electrification in the assessed economies. This success contributes to the strengthening of the social system dimension, which has the second highest median score among all seven ETRA dimensions.

- Developing Asia's median scores in labor rights and social protection cover indicators are lowest compared to the world and advanced economies' medians. As new jobs are created in a clean economy, strong labor rights and social protections will be crucial for developing Asian countries to ensure a just and equitable energy transition.

- As several developing Asian economies are fossil fuel-dependent, assessed countries also score lowest in vulnerability to climate disruptions and PM2.5 air pollution, representing long-term environmental and health consequences to their populations. This can strain health-care systems and reduce quality of life over the long term.

ETRA = energy transition readiness assessment, PM2.5 = particulate matter.

Note: Lower score = worse, higher score = better.

Source: Asian Development Bank.

Developing Asia accounts for more than half of the people in the world without access to clean cooking facilities.[26] Indonesia made significant progress in providing clean cooking (Box 9). Yet, more than half of the assessed economies score below the world median for clean cooking and heating access. Countries have leveraged diverse options to improve access to clean cooking, including liquefied petroleum gas, electric cookstoves, and biogas. Significant investment in developing and deploying innovative solutions is necessary to meet the target of universal access to affordable, clean cooking solutions by 2030.

Box 9

Clean Cooking Initiatives—Indonesia

The Government of Indonesia made cleaner cooking fuels more affordable by reducing kerosene subsidies and increasing liquefied petroleum gas (LPG) support. The government's flagship Kerosene-to-LPG Conversion Program (2007) replaced kerosene with LPG for cooking, distributing more than 50 million LPG stoves to households and small businesses. It reduced kerosene consumption by 95%, improving energy access and air quality.[a] In collaboration with the World Bank, the Indonesia Clean Stove Initiative (2012) promotes efficient biomass cookstoves for 40% of households relying on solid fuels. The program has spurred market development and awareness for cleaner cooking technologies in Indonesia.[b]

---

[a] *Energy Sector Management Assistance Program.* 2016. Indonesia Clean Cooking: ESMAP Supports Innovative Approaches to Build the Local Cookstoves Market, Helps Increase Access. 2 September.

[b] World Bank. 2014. Indonesia: Government Will Provide Universal Access to Clean Cooking Practices. Press release. 14 August.

Rising income levels and demographic shifts determine the shifting pattern of energy demand in developing Asia. Energy demand for transportation is expected to increase with rising levels of urbanization. The transportation energy mix is primarily fossil fuel-based. The PRC is leading globally in developing and adopting low-emissions transport, with deep technological and manufacturing expertise across the electric vehicle value chain. Governments in countries across developing Asia are incentivizing the adoption of electric vehicles and investing in the development of charging infrastructure. In addition to electrification, public transportation has the potential to accelerate mobility transformation. The median access to public transportation score in developing Asian countries is lower than that of the world, partly due to the spatial characteristics of its demography and population density. Expanding public transportation coverage can provide affordable and reliable service to commuters, supporting an equitable transition.

Similarly, developing Asian countries are considered "high risk" in terms of exposure to extreme temperatures, with a significant share of the population lacking access to cooling, including space cooling and refrigeration.[27] As climate change impacts intensify, access to sustainable cooling is essential for adaptation, which can further drive energy demand (Box 10).

---

[26] IEA. 2024. SDG7: Data and Projections—Access to Clean Cooking.

[27] Sustainable Energy for All. 2023. Regional Access to Cooling Risk. In *Chilling Prospects 2023*.

Box 10

Access to Sustainable Cooling

Developing Asia's rising demand for cooling must be met through clean energy sources. Higher demand for space cooling has emerged as a prevalent feature of developing Asia's energy landscape. Much of Asia's population is based in tropical latitudes and warm climates: in terms of cooling degree days and population, the world's 11 metropolitan areas with the highest cooling demands are all in Asia (India, Indonesia, Pakistan, the Philippines, Thailand, and Viet Nam).[a] As the incomes of Asia's middle classes rise, electricity access improves; and, as global temperatures continue to edge steadily upward, demand for and penetration of air conditioners (ACs) has increased dramatically. Globally—from 2000 to 2021—electricity demand for space cooling rose at an average of about 4% per year, twice as quickly as for water heating (footnote a). The AC market in developing Asia and the Pacific is projected to grow more than 6% annually to reach \$122 billion by 2030.[b]

Burgeoning electricity consumption to run ACs has major implications for developing Asia regarding overall and peak electricity demands, system reliability, and decarbonization goals. Indirect carbon dioxide (CO_2) emissions from space cooling nearly tripled during 1990–2022, reaching just over 1 gigaton of CO_2 in 2022.[c] Cooling's overall share of total electricity consumption grew from 8% to 19% in the same period, compounding the threat of supply constraints in many countries.[d]

A large share of the projected global growth in energy use for space cooling comes from developing Asia. The People's Republic of China (PRC), India, and Indonesia will contribute half the expected global increase by 2030. Nowhere has increased cooling electricity consumption been more pronounced than in the PRC. In 1990, cooling electricity consumption in the PRC was 6.6 terawatt-hours (TWh); by 2016, it was 450 TWh, a 68-fold increase. India's cooling-related energy demand is estimated to increase from 90 TWh in 2016 to 1,350 TWh by 2050, a 15-fold increase (footnote c). Indonesia is forecast to have a nearly 13-fold increase in cooling electricity demand to 340 TWh by 2050, an expansion which will likely see Indonesia running about 50% of all installed ACs in Southeast Asia.[e] Cooling electricity use in Southeast Asia's buildings is projected to rise about 275% over 2020 levels by 2040. Without stronger policies to encourage more efficient AC usage, member countries of the Association of Southeast Asian Nations will require an estimated 200 gigawatts of additional generation capacity by 2040, nearly double the region's installed capacity in 2017 (footnote e).

Other complex challenges exacerbate these growth dynamics. AC penetration remains low in the region's major countries, at only around 10% of the population in India, Indonesia, the Philippines, and Viet Nam; less than 30% in Thailand; and 60% in the PRC.[f] Many people in poor rural and urban areas remain at high risk due to a lack of access to cooling, with associated impacts on health and livelihoods. At the same time, building floor area has increased by about 60% since 2000 and is expected to rise by another 15% by 2030, most of which is being constructed in countries lacking comprehensive building energy codes.[g] Another significant issue is the imperative to phase out commonly used cooling refrigerant gases (hydrofluorocarbons) that are powerful greenhouse gases and exacerbate climate change.

Meeting cooling needs while ensuring a transition to a low-emissions future remains a serious and complex issue to address in the energy transition. Given expectations of a rapid increase in new appliance penetration in the coming years, developing Asia must avoid the lock-in of inefficient equipment that may take years to change out once in place. Ensuring that additional AC equipment to be installed is energy-efficient will be critical to reducing the impact of cooling on energy demand. The global average efficiency of ACs purchased by consumers has improved steadily in recent years.

continued on next page

Box 10 *continued*

Introducing more stringent minimum energy performance standards (MEPS) in the PRC—a major supplier of ACs across the region—has led to rapid market domination of the most efficient (Grade 1) ACs in that market. Variable-speed ACs—more efficient than fixed-speed models—have increased from 60% to 98% of the market share in the PRC since implementing the new MEPS.[h] These standards are projected to prevent at least 470 megatons of CO_2 emissions during 2020–2030, with further reductions possible through technological developments.[i] Enforcement of standards across Asia has not been consistent, meaning that cooling equipment performance across many Asian countries still has significant room for improvement.

To avoid the impacts of climate change, the average efficiency rating of new ACs needs to increase by at least 50% across all markets. Technological advancements include improved vapor compression cycles, low global-warming-potential refrigerants, demand response, and automated controls, which enhance energy efficiency and operational flexibility. Innovations—such as refrigerant-free or solid-state cooling units—are emerging.

The International Energy Agency's net-zero emissions scenario sets three primary cooling goals for 2030: (i) making 20% of existing building floor area and 100% of new constructions zero-carbon-ready, (ii) moderating AC temperature set points to 24°C–25°C, and (iii) enhancing the efficiency of new AC equipment to approach the best available. With these moves toward the best available products and improvements in the performance of the buildings in which they operate, estimated electricity demand for space cooling in buildings could be reduced by as much as 40% globally by 2030 (footnote i).

Several developing Asian countries have taken steps toward sustainable cooling. For example, India initiated its India Cooling Action Plan in 2019 to comprehensively plan for addressing cooling needs while mitigating climate impacts. Bangladesh and Cambodia published their National Cooling Plans in 2022, Indonesia released its plan in 2024, and Pakistan has announced its plans to do so by 2026. These plans aim to incorporate nature-based solutions and passive cooling techniques in urban development to meet housing demands and improve thermal comfort, while minimizing increases in cooling electricity demand.

Opportunities to scale up affordable space-cooling technologies also exist in housing schemes in South Asia. The Government of India's affordable housing scheme can incorporate climate-friendly insulation materials and passive cooling techniques, presenting a $1.5 trillion opportunity by 2040.[j] Similarly, Bangladesh and Pakistan have significant ongoing housing construction needs that can integrate sustainable cooling solutions.

In summary, ensuring equitable and sustainable access to cooling is critical for addressing urbanization and improving living standards. Advancing AC efficiency, implementing stringent MEPS, and adopting innovative technologies are essential to manage energy demand, reduce emissions, and enable the transition to clean energy generation. National and regional cooling action plans can also play a vital role in achieving these goals and mitigating climate impacts.

continued on next page

Box 10 *continued*

a   International Energy Agency (IEA). 2018. *The Future of Cooling – Opportunities for Energy-Efficient Air Conditioning.*

b   The Business Research Company. 2024. *HRO BPO Global Market Report 2024.*

c   IEA. 2023. *Energy System: Buildings—Space Cooling.*

d   IEA. 2023. *Space Cooling—Net Zero Emissions Guide.*

e   IEA. 2022. *Roadmap Towards Sustainable and Energy-Efficient Space Cooling in the Association of Southeast Asian Nations.*

f   IEA. Energy Systems: Buildings.

g   Collaborative Labelling and Appliance Standards Program. *Research Archive* (accessed 10 July 2024).

h   L. Zeng et al. 2023. *China's MEPS Lead to Major AC Market Transformation.* Collaborative Labeling and Appliance Standards Program (CLASP).

i   IEA. 2021. *Net Zero by 2050: A Roadmap for the Global Energy Sector.*

j   World Bank. 2022. *Climate Investment Opportunities in India's Cooling Sector.*

Developing Asia faces energy price volatility and transition risks, highlighting the need for targeted support and workforce readiness for a just and equitable transition. Expenditure of energy and energy services are a significant share of household expenditure, more so for households at the lower end of the income distribution. Aggregate figures for household electricity prices and cost of transport fuel in developing Asia indicate less severe affordability constraints than in advanced economies. However, aggregate values should be interpreted with caution, especially given the complex political economy of electricity and fuel tariffs, challenges in defining thresholds for energy poverty, and the sensitivities to individual needs and circumstances.

As the energy transition gathers pace, there can be potential short-term shocks to energy prices as the installed base of fossil-fuelled power plants is substituted with renewable alternatives. Economies in developing Asia are sensitive to global energy market volatilities, given the high levels of dependence on imported hydrocarbons. In addition to impacts on the affordability of energy, price volatilities in energy markets have cascading effects on prices of other essential goods and services, especially food and transportation, as evidenced in the global energy crisis following the Russian invasion of Ukraine. Low-income households demonstrate low elasticity of energy demand in response to price shocks, given the nondiscretionary nature of their energy consumption. Consequently, they are more vulnerable to adverse economic implications of energy market volatilities. Affordability constraints can drive consumers toward consuming inefficient and polluting forms of energy with environmental and health consequences and could undermine the social acceptance necessary for long-term climate goals. To provide timely and adequate support to households vulnerable to energy price shocks, countries must design effective emergency support mechanisms and closely monitor household energy consumption to target the affected population.

Universal social protection coverage is fundamental to cushion enterprises and workers exposed to adverse labor market implications of a net-zero transition. Workers, communities, and enterprises engaged in economic activities enabled by fossil fuels are exposed to transition risk through domestic climate policies and increasingly stringent environmental legislation in their export counterparts.

The share of the population covered by at least one social protection mechanism is low in assessed countries, except the PRC and select countries in Central Asia. Considering the relatively high number of migrant workers, informal workforce, and rising unemployment among youth, expanding coverage and adequacy of social protection is essential for the readiness of energy transition among developing Asian countries. Social protection measures must be developed with energy and labor market policies to ensure a just transition. If decarbonization efforts are accelerated, developing Asia will generate 2.9 million more jobs than if current policies are pursued, mainly in manufacturing solar and wind generation equipment, operation and maintenance of clean energy and smart grids, and construction. However, 1.4 million jobs are expected to be lost, with extreme losses in fossil fuel extraction like coal mining.[28] Moreover, a shift to renewables will create higher-skilled urban jobs, while rural, low-skilled workers may lose employment, necessitating reskilling efforts to help them transition to new occupations. Public services, training programs, and social protection measures should be implemented early to support affected workers and ensure smooth labor market transitions.[29]

3.6 Technology and Diffusion Ecosystem

Ability to deploy clean energy transition technologies such as solar photovoltaic (PV), wind power, biofuels, and electric vehicles—alongside enabling technologies like big data and artificial intelligence—will determine the speed and scale of the energy transition in developing Asia. Progress in these areas facilitates decarbonization and creates economic opportunities through enhanced efficiency and the emergence of new markets. According to the United Nations Trade and Development (UNCTAD), the market for green frontier technologies is projected to grow from $590.0 billion in 2020 to $2.1 trillion by 2030.[30] However, a country's ability to harness these opportunities and utilize technologies for the energy transition relies on a supportive environment for innovation, production, and diffusion of relevant technologies. The ETRA technology and diffusion ecosystem dimension encompasses a country's overall and energy-specific innovation and production capacity, access to skilled talent, and the spread of renewable and low-carbon technologies (Figure 14).

Developing Asia has massive potential for the adoption and production of clean energy technologies, but urgent efforts are needed to broaden and strengthen its technology and diffusion ecosystem. There has been a slowly improving—albeit uneven— landscape for clean energy adoption and production in developing Asia. The PRC is already a world leader in using, innovating, and producing clean energy technologies and services (Box 11). The PRC is the only assessed country to score above the advanced economy median score on the technology and diffusion ecosystem dimension, outperforming most countries in the number of clean energy patents, jobs in low-carbon industries, renewable energy capacity, and comparative advantage in the production of low-carbon technologies. This helps the PRC account for 60% of the world's manufacturing capacity for most mass-produced clean energy technologies and 40% of green hydrogen electrolyzer production (Box 12).[31]

---

[28] Footnote 5, p. 62.

[29] Footnote 27, p. 87.

[30] United Nations Conference for Trade and Development (UNCTAD). 2023. *Technology and Innovation Report 2023: Opening Green Windows—Technological Opportunities for a Low-Carbon World.*

[31] IEA. 2023. Clean Energy Supply Chains Vulnerabilities. In *Energy Technology Perspectives 2023*.

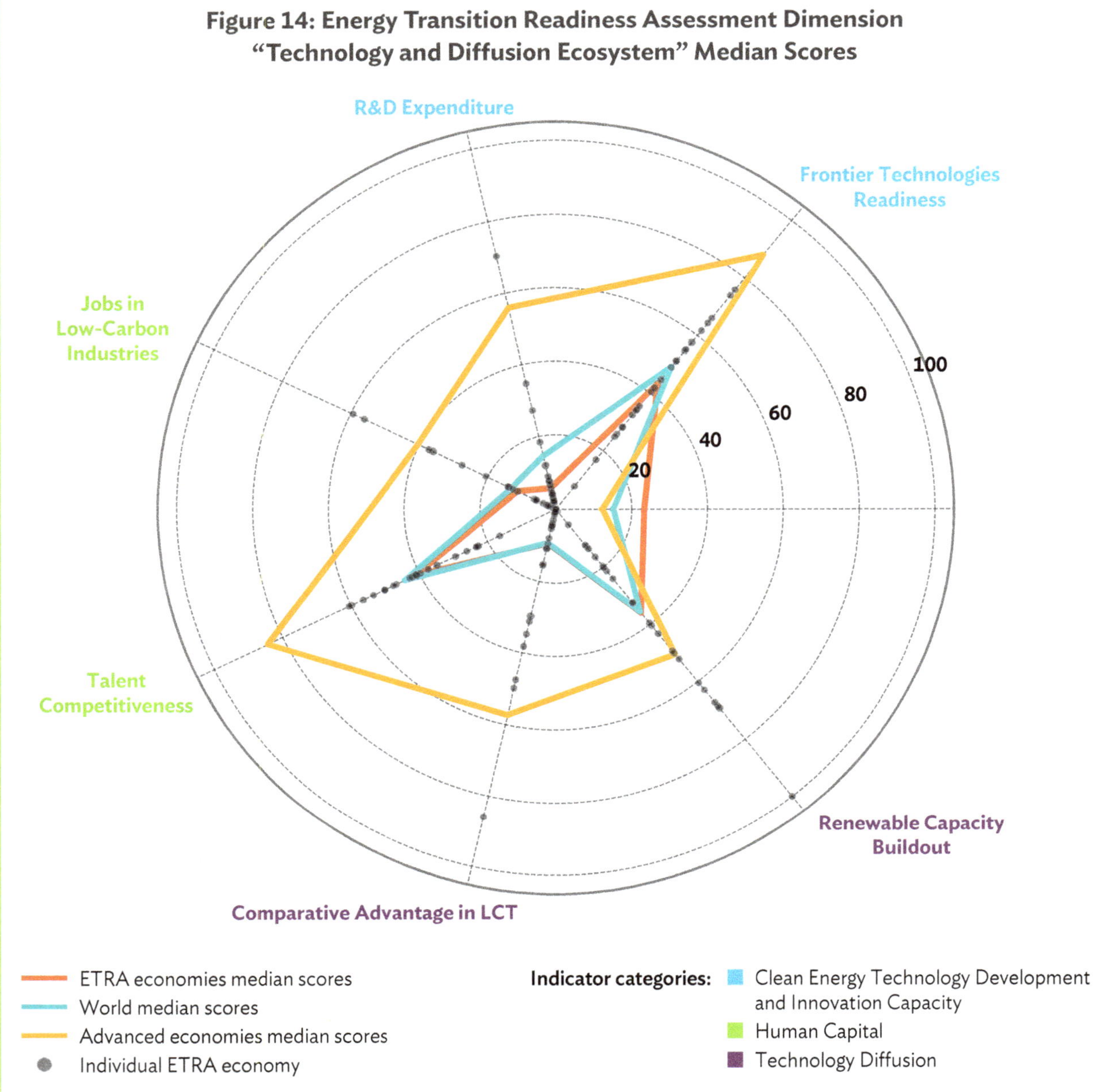

Figure 14: Energy Transition Readiness Assessment Dimension "Technology and Diffusion Ecosystem" Median Scores

Radar Plot Highlights

- Developing Asian economies score lowest for technology and diffusion ecosystem among all dimensions, which is also the dimension that has the largest score gap between developing Asian and advanced economies.

- Compared to the rest of the world, developing Asian economies score lower in R&D expenditure, frontier technologies readiness, and talent competitiveness, highlighting the region's lack of a skilled workforce, investments in R&D, and adoption of innovative technologies.

- A low median score for R&D expenditure correlates with low scores in comparative advantage in the adoption of LCT and jobs in low-carbon industries, as low investments in R&D expenditure can create challenges toward innovation and improving technologies as well as developing new industries and markets around LCT.

ETRA = energy transition readiness assessment, LCT = low-carbon technology, R&D = research and development.

Note: Lower score = worse, higher score = better.

Source: Asian Development Bank.

Box 11

Leadership in Clean Energy Research and Development —People's Republic of China

The People's Republic of China (PRC) has emerged as a global leader in clean energy research and development (R&D), driven by substantial investments, patent leadership, and cutting-edge innovations in solar, battery, and electric vehicle technologies. This success is built on strong collaboration among government bodies, private companies, and research institutions.

The PRC's public energy R&D budget was approximately $8.4 billion in 2019, with clean energy technologies receiving significant allocations. This represented about 0.06% of the PRC's gross domestic product, one of the highest rates globally. By 2023, the PRC's clean energy R&D investments surged further, supported by its five-year plans targeting advanced renewable technologies.[a] The PRC leads the world in green and low-carbon technology patents. In 2023, the country accounted for over half of the global total, with a 20% year-on-year increase in published green and low-carbon patent applications.[b]

R&D in solar photovoltaic, battery, and electric vehicle technologies. Leading companies such as LONGi Green Energy Technology Co. and Contemporary Amperex Technology Co. Limited have propelled the PRC to the forefront of renewable energy R&D. LONGi has been instrumental in advancing solar photovoltaic technologies, setting a world record for crystalline silicon solar module efficiency at 25.4% in 2024.[c] The PRC's BYD Co. Ltd. has significantly increased its R&D expenditures, reaching about CNY39.6 billion ($5.47 billion) in 2023, marking a 112% year-on-year growth. This investment focuses on advancements in battery chemistry, integrated powertrains, and electric vehicle platforms.[d]

Universities and research labs. Leading universities such as Tsinghua University and Zhejiang University have spearheaded renewable energy R&D projects, supported by major industrial partnerships. The National Institute of Clean-and-Low-Carbon Energy conducts advanced research in hydrogen energy, carbon capture, and next-generation power electronics (footnote a).

[a] International Energy Agency. 2022. *Tracking Clean Energy Innovation: Focus on China.*

[b] *Xinhua.* 2024. China Leads Global Green Patent Applications with Over 50 Pct Share. *China Daily.* 7 August.

[c] V. Shaw. 2024. Longi Green Energy Sets World Record for Solar Module Efficiency at 25.4%. *PV Magazine.* 24 October.

[d] *Xinhua.* 2024. Economic Watch: China's clean energy sectors gain edge via innovation, supply chain. 30 April.

Beyond the PRC, nine developing Asian economies—including most Southeast Asian states and India— score above the world median for this dimension. Despite the need for further improvements, these countries have a large and increasingly competitive industrial workforce, clean energy-related mineral deposits, and growing manufacturing capabilities in clean energy and ancillary industries, including automotive and electronic equipment.[32] With rising GDP and increasing population, these states are major markets for clean energy.

---

[32] C. Y. Park and A. C. Melendez. 2024. Building Resilient and Responsible Critical Minerals Supply Chains for the Clean Energy Transition. *ADB Briefs.* No. 298. p. 16.

Box 12

Green Hydrogen

Developing Asia's climate goals require rapidly decarbonizing hard-to-abate industries and all other sectors of these economies. Energy efficiency, electrification, and renewables can achieve 70% of the mitigation needed. The use of low or zero-carbon hydrogen can meet an additional 10% of mitigation needs by decarbonizing end uses such as heavy industry, long-haul transport, and seasonal energy storage.[a]

About 100 large-scale hydrogen project proposals were announced by the end of 2023. An estimated $27 billion in direct investment through 2030 has reached a final investment decision, of which about 18% are in the People's Republic of China.[b]

Other developing Asian economies have also announced ambitious green hydrogen plans. In January 2023, the Government of India launched the revised National Green Hydrogen Mission with an outlay of about $2.41 billion, with an expected production capacity of at least 5 million metric tons per annum (MMTPA) by 2030. The 2024 hydrogen demand of about 6 MMTPA is expected to increase 2.5–3.5 times by 2040. In partnership with the International Solar Alliance, the Asian Development Bank (ADB) supported the development of the Green Hydrogen Innovation Centre.[c]

Bangladesh launched a pilot hydrogen production project in 2021; the Government of Bhutan published the Hydrogen Roadmap in 2024; Sri Lanka worked toward developing the National Hydrogen Roadmap in 2023; and Nepal organized a green hydrogen summit in 2022, in collaboration with Global Green Growth Institute.

Indonesia, Malaysia, Singapore, and Viet Nam also issued official national green or low-carbon hydrogen supply strategies. Malaysia's Hydrogen Economy and Technology Roadmap of 2023 discusses opportunities being implemented or readied for implementation.[d] Indonesia's hydrogen strategy focuses on the transport and hard-to-abate industries and gradually transitions to green hydrogen to create a hydrogen economy.

Drivers and barriers to green hydrogen deployment. Cost parity between green and gray hydrogen, electrolyzer costs and performance, and government support and incentives for green and low-carbon hydrogen are primary drivers for green hydrogen take-off.

Despite the growing interest in green hydrogen, a mix of regulatory, technological, and financing barriers and limited market uptake remain significant constraining factors. Investments in green and low-emission hydrogen projects remain low and are inconsistent with the net-zero target by 2050. Other prominent barriers are comparatively high costs of renewable electricity (which accounts for 70%–80% of the green hydrogen production costs), high capital costs and low efficiencies of electrolyzers, high costs of transportation and storage of hydrogen, an insufficiently supportive regulatory environment, and limited availability of commercial and concessionary financing.

Establishing a green hydrogen industry at scale at 2024 production costs would require vast concessionary financing and grant assistance from multilateral and bilateral sources. As such, developed country governments and enterprises do more to address technology risks and bring down costs before they commit to any sizable investment.

continued on next page

Box 12 *continued*

Electrolyzer technologies. In 2023, the installed electrolysis capacity worldwide was 1,400 megawatts, with an estimated 5,200 megawatts installed by the end of 2024.[e] However, more than 200 gigawatts of electrolysis capacity is needed by 2030 to remain on track for the 2050 net-zero target. Alkaline electrolyzers are already at the commercial stage and have lower investment costs than other technologies. They benefit from a simple system design and have other applications in the chemical industry that leads to the presence of an existing supply chain that can be scaled up. The proton exchange membrane electrolyzer is better suited for variable electricity supply, has the smallest plant footprint (kilograms of hydrogen per square centimeter), and can produce the highest hydrogen purity. About 40% of total electrolyzer capacity is installed in the People's Republic of China.[e]

Opportunities for green hydrogen in developing Asia. Green hydrogen opportunities do exist in the region. ADB has made a small beginning and is focused on (i) supporting policy, strategy, and road map development; (ii) engaging with the industry on standards and knowledge sharing; (iii) piloting green hydrogen technologies and business models for demonstration and scale-up; and (iv) financing green hydrogen production, transport, and distribution projects. ADB could also identify industrial enterprises with significant green hydrogen demand and sufficient space on-site or nearby for locating demonstration plant(s). The viability of the green hydrogen electrolyzers will be enhanced if such enterprises can off-take surplus and inexpensive off-peak electricity from large hydropower plants.

[a] United Nations Climate Change. 2024. Action to Decarbonize High Emitting Industrial Sectors Gathers Pace – Key Steps by the Technology Executive Committee. News Article. 6 May.
[b] Hydrogen Council. 2023. *Hydrogen Insights 2023*.
[c] Asia Clean Energy Forum. 2024. *Spotlight Session—Green Hydrogen: Its Development, Status and Prospects*. ADB Headquarters, Manila. 15 June.
[d] M. N. Arianto et al. 2024. Pushing Indonesia's Decarbonisation Agenda Through Low-Carbon Hydrogen National Strategy. *ASEAN Center for Energy*. 26 June.
[e] International Energy Agency. 2024. *Global Hydrogen Review 2024*.

India, Malaysia, Thailand, and Viet Nam score above the global median for comparative advantage in producing LCT, with India, Thailand, and Viet Nam improving their comparative advantage since 2013. According to some estimates, Southeast Asia's solar PV, battery, and electric two-wheeler markets could generate $90 billion–$100 billion in revenue by 2030, creating up to 6 million jobs by 2050 (footnote 15).

However, despite progress, the technology and diffusion ecosystem is the lowest scoring dimension for assessed countries, with a particularly large gap compared to advanced economies. Generally, low scores in research and development (R&D) spending, the readiness of frontier technologies, and talent competitiveness reveal that many developing Asian countries lack the R&D financial support, skilled workforce, and other enablers necessary for adopting and developing clean energy technologies on the necessary scale. Assessed countries' challenges in encouraging local clean energy innovation and production ecosystems in the face of advanced economy competition are further reflected in typically below advanced economy scores for the number of jobs and comparative advantage in LCT products.

The former reinforces gaps in the formation and access to a workforce required for the energy transition, while the latter highlights that while some countries have made progress, most assessed economies still have a relative disadvantage in the production and export of energy transition-related goods. As a result, much of the push that led to technological advances and economies of scale in clean energy technologies has occurred in advanced economies and the PRC.[33]

In the short to medium term, increasing the diffusion and production of already developed clean energy technologies will likely significantly impact decarbonization efforts. For emerging and developing economies (EMDEs), more than three-quarters of clean energy investment needs during 2024–2035 are in commercially proven technologies like utility-scale wind and solar, modernized grids, and electric mobility.[34] This requires creating supportive policies and regulations that reduce investment barriers, improve access to credit, and provide incentives for private-sector investment in clean energy equipment.

Promoting international trade, foreign direct investment (FDI), and technology licensing can facilitate the transfer of technologies from more developed markets and help build local production capacity.[35] EMDEs introducing climate policies and lower tariffs see a greater rise in LCT imports and higher green FDI inflows.[36] Trade is also critical for access to the most affordable technologies, with the price of solar PV modules and electric vehicle batteries dropping by more than half during 2015–2022 due to increased production and scaling in the PRC in part.[37]

Creating demand for clean energy is also crucial. Policymakers can promote the adoption of clean energy by offering incentives like tax credits, subsidies, or feed-in tariffs for clean energy projects and products. Governments can also establish ambitious renewable energy targets, create credible carbon pricing mechanisms, and reduce fossil fuel subsidies to encourage the shift to greener alternatives. Policies to make clean energy prices competitive with fossil fuels will be particularly important, given lower income levels in developing economies. The International Monetary Fund estimates that a major increase in climate policies can boost green patent filings by 10% in 5 years and that EMDEs that introduce climate policies and lower tariffs see a stronger rise in LCT imports and higher green FDI inflows.[38]

The region's growing industrial and economic weight also requires increased efforts in developing and employing technologies for hard-to-abate manufacturing (steel, aluminum, cement, and chemicals) and transportation (aviation, shipping, trucking) sectors, including commercial scaling of carbon capture, utilization, and storage (CCUS) technology and green hydrogen (Box 12). According to the World Economic Forum's *Net Zero Tracker 2024,* almost half of the required emission reductions of hard-to-abate sectors require the use of technologies that are currently commercially unavailable. As a result, commercial scaling of CCUS and green hydrogen technologies will be necessary to address decarbonization efforts for hard-to-abate sectors in developing Asia.[39]

---

[33]   IEA and International Finance Corporation (IFC). 2023. *Scaling Up Private Finance for Clean Energy in Emerging and Developing Economies.* p. 32.

[34]   IEA. 2024. *Reducing the Cost of Capital: Strategies to Unlock Clean Energy Investment in Emerging and Developing Economies.* pp. 17–18.

[35]   Footnote 33, pp. 31–33.

[36]   Z. Hasna et al. 2023. Green Innovation and Diffusion: Policies to Accelerate Them and Expected Impact on Macroeconomic and Firm-Level Performance. *Staff Discussion Notes.* No. SDN/2023/008. International Monetary Fund.

[37]   Footnote 33, p. 32.

[38]   Z. Hasna, F. Jaumotte, and S. Pienknagura. 2023. How Green Innovation Can Stimulate Economies and Curb Emissions. *IMF Blog.* 6 November.

[39]   World Economic Forum. 2024. Net-Zero Industry Tracker 2024. p. 5.

Within this context, industrial clusters will play an important role in accelerating diffusion by acting as centers for scaling technology, risk-sharing, optimization resourcing, and aggregating demand for clean energy. For instance, Indonesia's Jababeka Net Zero Industrial Cluster—which includes more than 1,800 companies—has developed a decarbonization road map and created partnerships for shared infrastructure to enhance the circularity of industrial solid waste and wastewater management. In the PRC, the Tianjin Economic–Technological Development Area—which includes sectors such as aviation, oil and gas, chemicals, and power generation—is focused on scaling the integration of clean technologies such as CCUS and renewables. This includes implementing a geothermal-well heat supply project and expanding solar PV power generation capacity.[40]

Lastly, as chapter 3.5 (the social system dimension analysis) mentions the shift to clean energy presents a significant opportunity for developing Asian countries to create millions of new jobs in sectors like solar and wind manufacturing, clean energy maintenance, and smart-grid operations. To ensure that the workforce is effectively skilled, upskilled, and reskilled, countries should develop targeted training programs, align educational curricula with clean energy needs, and provide reskilling opportunities for workers transitioning from fossil fuel sectors. Partnerships between governments, academia, and industry stakeholders can help design training programs that meet market demands, while initiatives like vocational training, on-the-job mentorship, and certification can support workers in acquiring new skills for emerging clean energy roles.[41]

3.7 Macroeconomic and Investment Environment

Strong macroeconomic fundamentals and favorable investment conditions are critical for attracting and allocating diverse domestic and foreign investments for capital-intensive clean energy projects and other transition needs. The ETRA macroeconomic and investment environment dimension assesses a country's monetary and fiscal conditions, financial stability and risk, and investment climate for the clean energy transition (Figure 15).

Developing Asia witnessed positive—albeit concentrated—momentum in investments in the clean energy transition. However, greater and broader investment is required to meet future energy needs and decarbonization goals. Since 2013, clean energy investment in developing Asia has grown more than 900%, reaching $729.4 billion in 2023, accounting for 45% of global investment (Figure 16). Moreover, the 3-year moving average for investment in clean energy as a share of GDP has increased for 15 assessed economies over the same period, reflecting clean energy's growing economic role and shift toward effective decarbonization. Nevertheless, transition and climate finance are highly concentrated in the PRC, which accounted for nearly 90% of the region's investment during 2013–2023. On the other hand, most other developing Asian countries lag and score below the world median for clean energy investment as a share of GDP. This indicates the need to broaden transition funding throughout the region. Developing Asian economies must increase annual clean energy investment (electricity supply and energy end use) to $1.4 trillion to meet energy and decarbonization needs under the accelerated global net zero scenario during 2020–2050.[42] However, the median dimension scores of developing Asian countries remain below world levels. This reflects the typically challenging macroeconomic conditions and investment climate faced by developing Asian economies. Half of the 26 developing Asian countries score below the world median, and only three (the PRC, Malaysia, and Viet Nam) score above the median score of advanced economies.

---

[40] World Economic Forum. 2024. *Transitioning Industrial Clusters: Annual Report.*

[41] IEA. 2022. *Skills Development and Inclusivity for Clean Energy Transitions.*

[42] Footnote 5, p. 83.

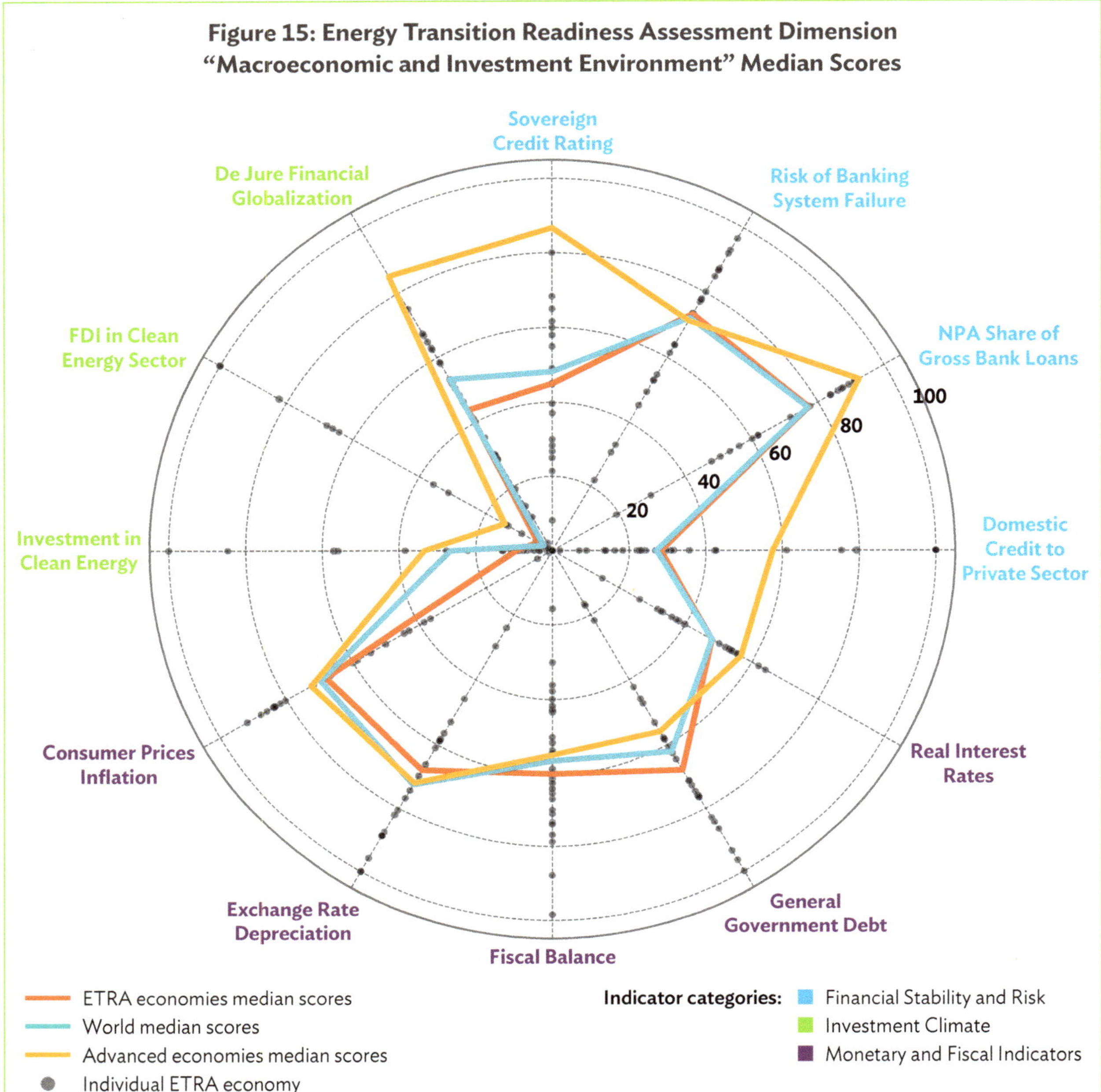

Radar Plot Highlights

- Median dimension scores of developing Asia remain below world levels, reflecting typically challenging macroeconomic conditions and investment climate.

- Especially compared to more economically stable advanced economies, developing Asia's macroeconomic uncertainty, risk, and higher cost of capital are reflected in lower scores for real interest rates, domestic credit to the private sector, and sovereign credit ratings. Low scores for de jure financial globalization and investment in clean energy indicators reflect opportunities for increased sector funding and improving financial opportunities for foreign investors.

- Developing Asian economies have maintained lower government debt levels, underscoring some room for public sector funding for clean energy.

ETRA = energy transition readiness assessment, FDI = foreign direct investments, NPA = nonperforming asset.

Note: Lower score = worse, higher score = better.

Source: Asian Development Bank.

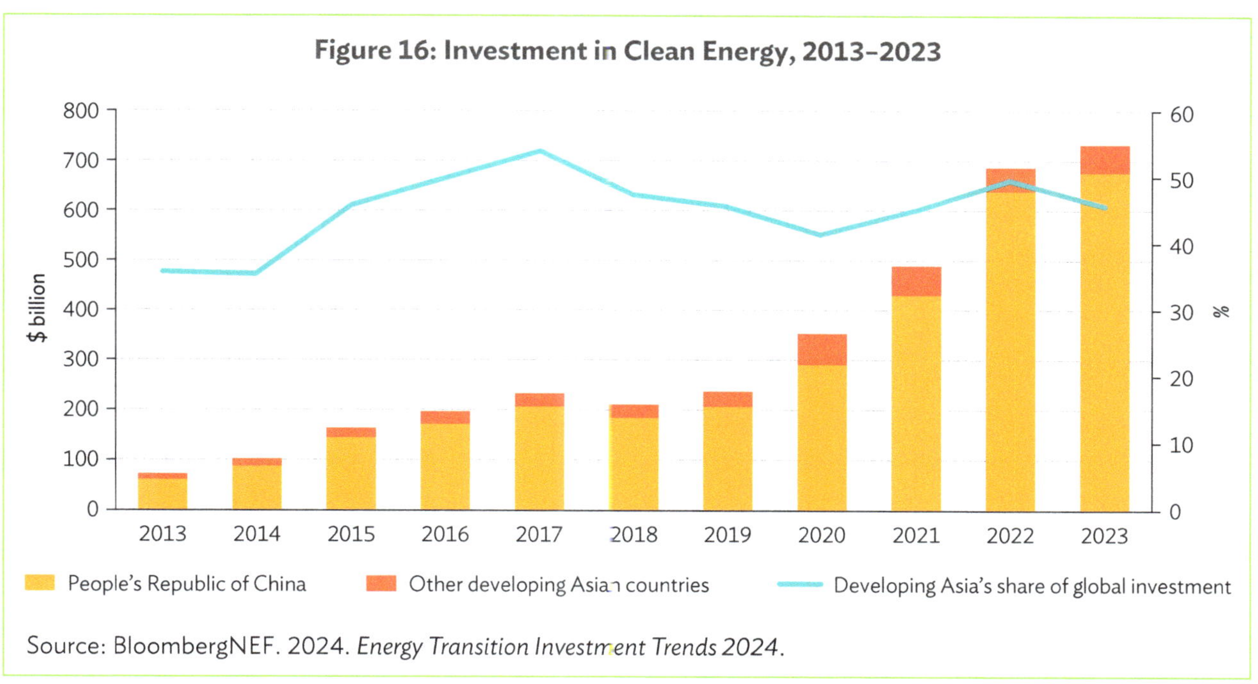

Figure 16: Investment in Clean Energy, 2013–2023

Source: BloombergNEF. 2024. *Energy Transition Investment Trends 2024.*

Challenging macroeconomic conditions hinder greater funding for clean energy in most developing Asian economies. Median ETRA assessed country scores for the macroeconomic and investment environment dimension remain below global levels, with only the PRC, Malaysia, and Viet Nam outperforming the advanced economy levels. Many assessed economies face higher inflation, exchange rate volatility, and lower sovereign credit ratings. This reflects the economic uncertainty that feeds the real and perceived risk investors associated with developing countries. Combined with recent years' interest rate hikes, these factors drive up the already high cost of capital for many developing Asian countries. For instance, in assessed economies, median real interest rates over the 3 years to 2023 are more than 10 times higher than the levels for advanced economies. This financing cost disparity directly translates into the often more capital-intensive and riskier clean energy projects, with the expected equity internal rate of return on solar utility-scale PV projects in some developing countries being double those in Europe or the United States.[43]

Underdeveloped domestic financial markets further limit private investment, with domestic credit to the private sector below average for many regional states. Simultaneously, while government debt-to-GDP is generally below global levels, indicating some room for further public sector funding, elevated debt ratios, and fiscal deficits since the coronavirus disease (COVID-19) pandemic, coupled with low credit ratings and higher interest rates, restrict the public sector's ability to accelerate clean energy investments and make up for lower private financing in many countries. This is of particular concern for developing Asia, given that public sources of finance, such as state-owned enterprises, account for nearly 60% of clean energy investments.[44]

---

43   IEA. 2024. *World Energy Investment 2024.* p. 28.
44   Imperial College London and IEA. 2023. *ASEAN Renewables: Opportunities and Challenges.* p. 8.

Developing Asian countries must prioritize creating more conducive and less risky environments for private sector investment in clean energy. Given their investment needs and public sector constraints, in the coming years, developing Asian economies must prioritize a shift toward a private sector-led financing model to meet energy transition needs. According to the International Energy Agency, more than 60% of annual clean energy investment during 2026–2035 must come from private finance for emerging and developing economies (EMDEs) to reach net zero by 2050.[45] To mobilize this private investment, developing Asian countries must prioritize mitigating investment risks and making projects more bankable, with grants, guarantees, low-interest loans, and blended finance from sources such as development finance institutions and philanthropies and by leveraging new financing mechanisms.[46] The International Finance Corporation estimates that India, the Association of Southeast Asian Nations (ASEAN), and other Asian EMDEs (excluding the PRC and finance needs for public investments) will require $23 billion annually in concessional funds during 2026–2030 to help leverage more than $300 billion in private funding annually (footnote 49). Project aggregation platforms and securitization vehicles can further facilitate investment by pooling smaller, riskier energy projects into standardized, diversified, investment-grade portfolios, reducing costs and spreading risks.[47] Since currency risks are a major driver of the cost of capital, blended financing and global facilities like the Currency Exchange Fund are also important in expanding access to and reducing currency hedging costs.[48] Energy transition mechanisms are also emerging as new funding tools that allow investors to buy and retire carbon-heavy assets, replacing them with renewable energy. This is vital given developing Asia's dependence on coal power (Box 3).

Developing Asia must also act to access the $7 trillion global sustainable financial market. The issuance of green, sustainability, and sustainability-linked bonds and transition loans is increasingly crucial for developing Asian economies.[49] During 2020–2022, the annual issuance of these debt products in developing Asian economies increased from $70 billion to $295 billion, with 89% of the financing coming from the private sector in 2022. However, sustainable bonds only accounted for 1.8% of the region's bonds that same year, reflecting the potential for greater use of these debt instruments.[50] Developing industry guidelines, reporting frameworks, taxonomies, and robust and credible third-party certifications will be vital for tapping into this market further. The ASEAN Green Bond Standard and ASEAN Taxonomy exemplify multilateral efforts to standardize sustainable financial product standards (Box 13). Indonesia, Malaysia, and Thailand are examples of countries that have developed their taxonomies to support this market expansion further.[51]

Moreover, to attract greater private investment in clean energy, policies must ensure transparency and predictability (which help reduce regulatory risk) and open their utility sectors by allowing independent power producers, using standardized power purchase agreements, and holding transparent auctions.

---

[45]   Footnote 33, p. 129.

[46]   Blended finance combines concessional donor funds with commercial funds from private investors, multilateral development banks, and others to de-risk financing, making clean energy projects more attractive.

[47]   Footnote 33, p. 145; and B. Handler, M. Hayes, and M. Bazilian. 2021. 5 Ways to Boost Renewable Energy Investment in Developing Economies. *World Economic Forum.* 29 June.

[48]   Footnote 34, p. 23.

[49]   Green bonds raise funds specifically for projects with environmental benefits, sustainability bonds are used to finance projects with environmental and social benefits, and sustainability-linked bonds tie the cost of financing to key sustainability performance indicators. UNCTAD. 2024. *2024 World Investment Report.*

[50]   Footnote 5, p. 84.

[51]   P. Janssen. 2024. South-East Asia Poised for Sustainable Bond Lift-Off. *The Banker.* 29 April.

Box 13

Green, Social, Sustainable, and Other Labeled Bonds Market in ASEAN

Several Association of Southeast Asian Nations (ASEAN) countries are among the top emitters in the world, making them vulnerable to climate change-induced shocks. According to an Asian Development Bank estimate, ASEAN countries will need $210 billion annually during 2016–2030—equivalent to 5.7% of the region's gross domestic product—in climate-resilient infrastructure.

The sustainable bond market in ASEAN began to develop following the introduction of the ASEAN Green (2017), Social (2018), Sustainability (2018), and Sustainability-Linked (2021) Bonds Standards—a set of labels collectively referred to as GSS+ Bonds. The GSS+ bonds market expanded with more than $50 billion worth of bonds issued in ASEAN member countries. A unique feature of the GSS+ bonds is the emergence of local currency-dominated bond markets in ASEAN+5, which comprise Indonesia, Malaysia, the Philippines, Singapore, and Thailand. As of June 2024, about 71% of GSS+ bonds were issued in local currency in the ASEAN+5 bonds market.

Source: Asian Development Bank. 2024. *Mobilizing Private Capital for Sustainable Development through Sustainable Bonds: Case Studies from Indonesia and Thailand.*

Clean energy incentives—including a multiyear energy strategy with short-term renewable targets, carbon pricing mechanisms (Box 5), and carbon removal governance—create market certainty and demand for clean energy investment. Additionally, broad business-friendly measures like favorable tax policies, permitting foreign direct investment, streamlined permitting, and profit repatriation in foreign currency are essential for renewable energy investment and reducing risks for foreign investors.[52] For instance, while many assessed developing Asian countries are slowly becoming more open to foreign investment, they still score below the advanced economies' median for de jure financial globalization, indicating that financial openness remains a common barrier to foreign direct investment.

Developing Asian economies also need to deepen local financial systems and capital markets. Domestic private investment has been a major driver of clean energy investment in larger economies like the PRC and India.[53] Domestic banks are vital in financing smaller climate-related projects, such as access to electric vehicles and solar energy for small and medium-sized enterprises and households. Developing local bond, equity, and derivatives markets is essential for larger projects to mobilize more local currency and reduce reliance on foreign capital. Multilateral development banks and other international players are essential in strengthening local banks by offering de-risking tools and capacity-building programs.[54]

---

[52] B. Handler, M. Hayes, and M. Bazilian. 2021. 5 Ways to Boost Renewable Energy Investment in Developing Economies. *WEF.* 29 June.

[53] Footnote 33, p. 125.

[54] Footnote 33, pp. 147–148.

3.8 Regulatory Environment

A country's regulatory and policy environment supports a successful energy transition. A comprehensive regulatory framework—encompassing laws, policies, and regulations across all sectors—helps align stakeholder interests; enhances governance; and sets incentives and rules that guide clean energy investment, innovation, and infrastructure development. Conducive regulatory conditions help reduce uncertainty and risk and streamline government processes. Well-thought-out, long-term policies are critical for ensuring energy transition security, resilience, and equity.

The ETRA regulatory environment dimension assesses a country's regulatory readiness to advance climate commitments, comprehensive energy policy frameworks, and governance quality (Figure 17).

The median regulatory environment dimension score for the assessed economies exceeds the world median. Sixteen out of 26 assessed countries scored above the world median, indicating that most are improving their regulatory frameworks. Georgia, Malaysia, and Thailand are notable performers, with median scores among the closest to the advanced economies' medians. However, the other assessed economies are lagging. This underscores the need for continued efforts to strengthen regulatory frameworks, improve governance, and enhance policy environments to support the clean energy transition in developing Asian countries, such as Viet Nam's renewable energy expansion (Box 14).

Box 14

Leadership in Renewable Energy amid Grid Stability Concerns—Viet Nam

Viet Nam has established itself as a leader in renewable energy, which accounts for 19% of the total electricity coming from renewable energy. The Power Development Plan VIII sets an ambitious target for renewables to constitute 32% of electricity generation by 2030. The country excels in the World Bank Regulatory Indicators for Sustainable Energy 2021, scoring 84/100 in renewable energy, with strong legal frameworks and planning mechanisms for renewable energy expansion.[a]

High feed-in tariffs for solar and wind have driven exponential growth. Viet Nam's renewable energy generation rose from about 997 gigawatt-hours in 2018 to 37,865 gigawatt-hours in 2022, making Viet Nam Southeast Asia's renewable energy leader.[b]

However, the rapid surge in renewable energy created safety concerns for the national power grid, particularly in the central region rendering the EVN (Viet Nam's state-owned electric utility company) to curtail the renewable energy electricity offtake and consequently inflicting financial losses for power project developers.[b]

[a] Energy Sector Management Assistance Program. 2022. *Regulatory Indicators for Sustainable Energy (RISE) 2022: Building Resilience*. World Bank. p. 40.

[b] L. H. Hiep. 2024. The Unexpected Twist in Vietnam's Renewable Energy Saga. *Fulcrum*. 4 January.

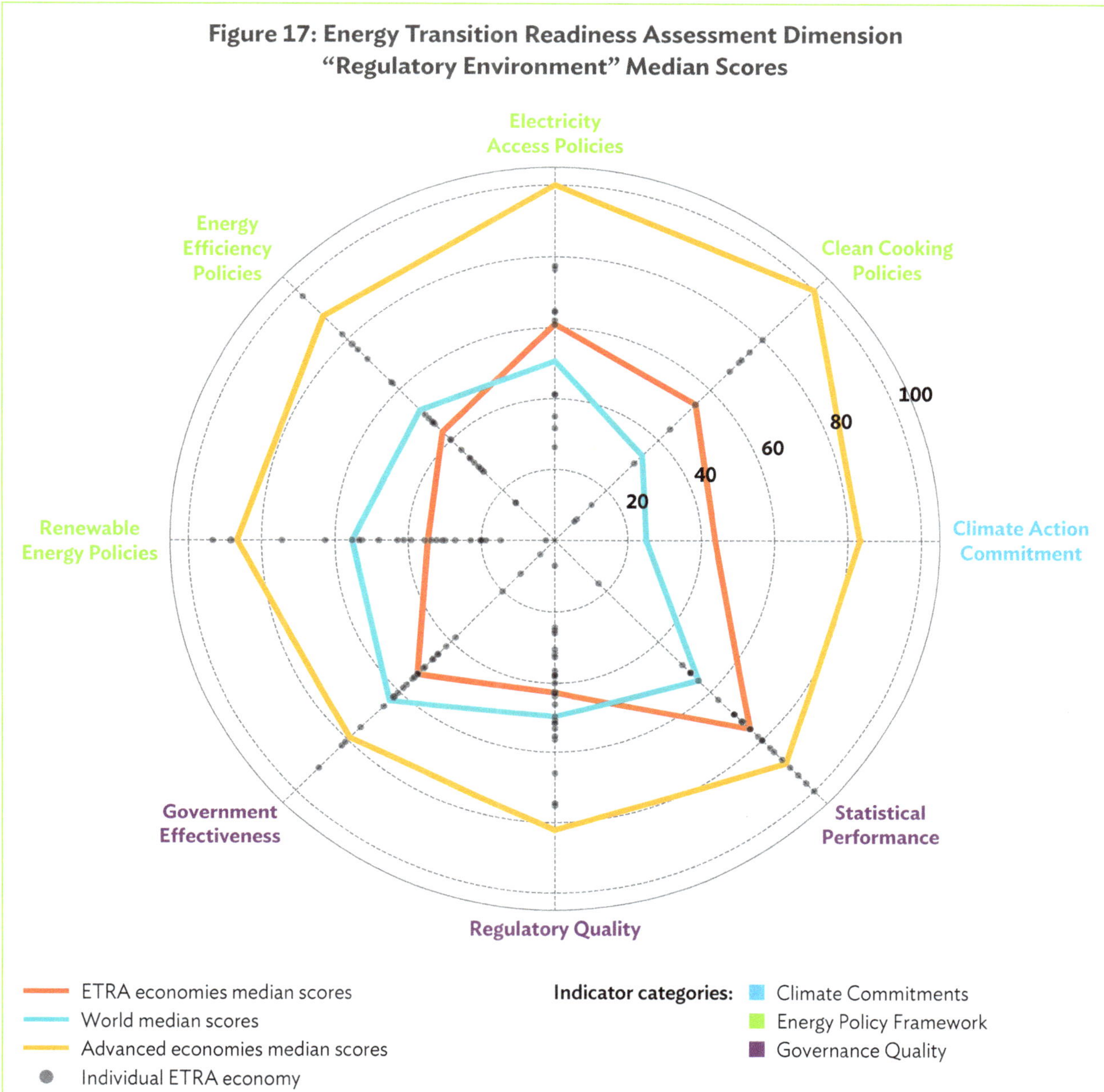

Figure 17: Energy Transition Readiness Assessment Dimension "Regulatory Environment" Median Scores

ETRA economies median scores
World median scores
Advanced economies median scores
Individual ETRA economy

Indicator categories: Climate Commitments
Energy Policy Framework
Governance Quality

Radar Plot Highlights

- Assessed economies lag advanced economies in terms of establishing long-term decarbonization strategies, stringent climate regulations, and credible energy transition efforts, as reflected by the lower median score in climate action commitment compared to that of advanced economies.

- Developing Asian countries demonstrate a higher score in electricity access and clean cooking policies, showing government priorities during 2005–2024. However, they lag in renewable energy and energy efficiency policies, highlighting the need to develop a more comprehensive energy policy framework.

- While developing Asian countries have lower scores in regulatory quality and government effectiveness, they maintain strengths in statistical performance. Leveraging these statistical capabilities for better data-driven policymaking and monitoring can increase readiness for the energy transition.

ETRA = energy transition readiness assessment.

Note: Lower score = worse, higher score = better.

Source: Asian Development Bank.

Developing Asian countries must develop long-term climate strategies. While Bhutan, Georgia, and Sri Lanka are on par with advanced economies' median scores for commitment to climate actions, most developing Asian countries have gaps in the climate policy landscape. Most developing Asian countries have yet to present their second nationally determined contribution (NDC), develop related long-term strategies, or enact climate framework laws (Figure 18).[55] As a result, many developing Asian countries' commitment to long-term decarbonization policy may lack credibility and clear energy transition pathways, thereby creating regulatory uncertainty.

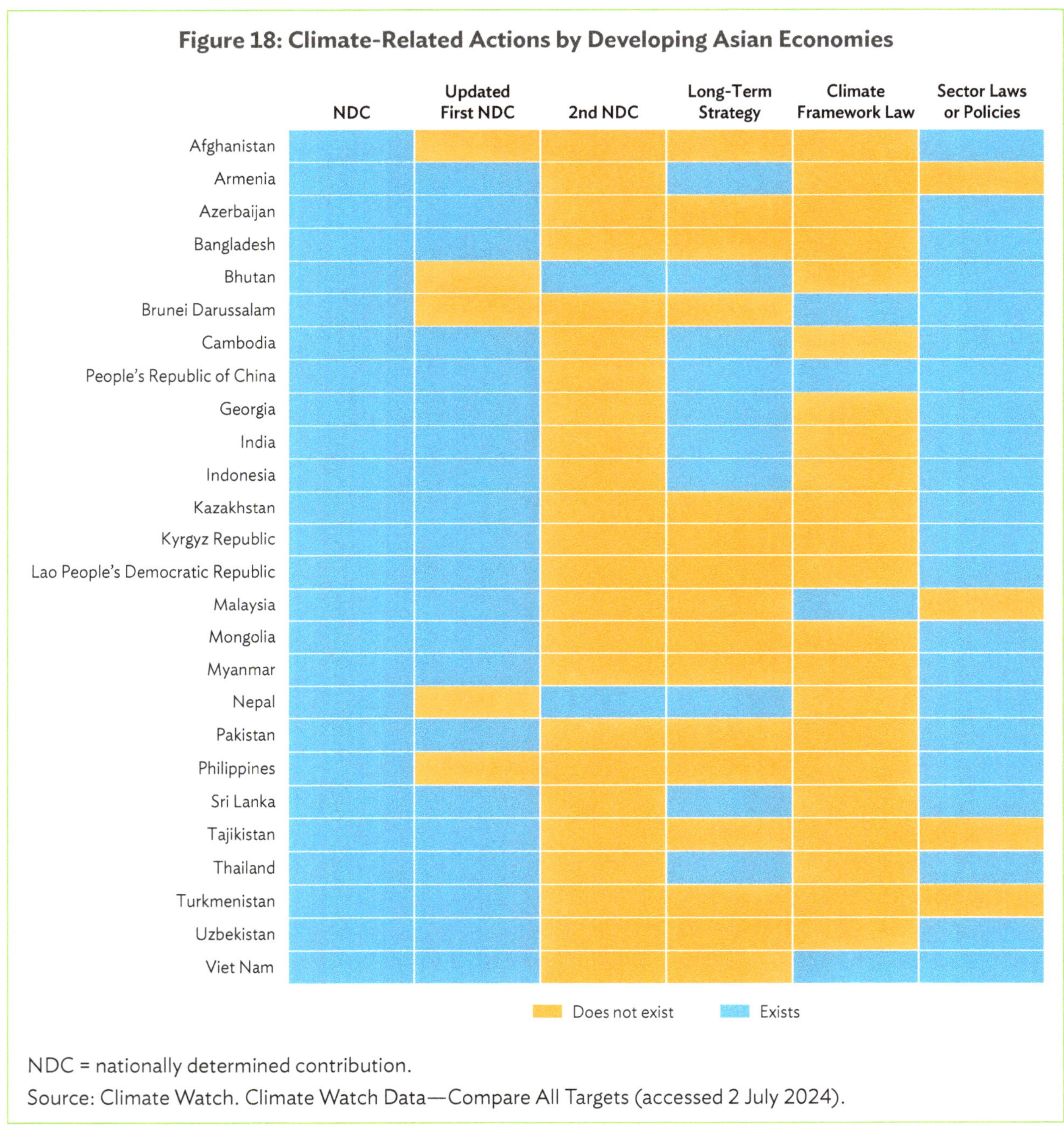

NDC = nationally determined contribution.
Source: Climate Watch. Climate Watch Data—Compare All Targets (accessed 2 July 2024).

---

55   Climate Watch. Climate Watch Data—Compare All Targets (accessed 2 July 2024).

Diverse regulatory landscapes shape the way toward clean energy solutions, noting path dependence and developing Asia's need to balance energy transition priorities with outstanding development needs. Many assessed economies lag in effectively supporting renewable energy, with their median scores for both renewable energy policies and energy efficiency policies trailing the world and advanced economies. However, large developing Asian economies like the PRC, India, and Viet Nam have created a strong policy environment that encourages renewable energy deployment, builds green energy corridors, and allows industry to procure renewable energy and time-of-day tariffs directly.[56]

Developing Asian economies must strengthen energy efficiency policies and regulatory frameworks to address technical and commercial losses across the energy system. While energy efficiency policies in the assessed economies span a wide range—such as building codes, appliance standards, and industrial efficiency programs—they must create enabling environments that drive the adoption of best-in-class technologies and new business models that accelerate energy efficiency across their economies. The PRC and India have led the implementation of diverse policy measures to enhance energy efficiency (Box 15).[57]

Box 15

Leadership in Energy Efficiency—People's Republic of China

The People's Republic of China (PRC) has emerged as a global leader in energy efficiency by implementing comprehensive policies, setting ambitious targets, and ensuring consistent follow-through in their execution. Notable policy and planning initiatives include (i) the 14th Five-Year Plan (2021–2025), setting a target of a 13.5% reduction in energy intensity below 2020 levels by 2025;[a] (ii) Dual Control Mechanism—introduced in the 13th Five-Year Plan, this mechanism sets provincial targets for energy intensity reduction and total energy consumption;[b] (iii) sector initiatives, which implement benchmarks for energy-intensive industries, efficiency standards for coal-fired power plants, building energy codes, and appliance standards—e.g., the 2024–2025 Action Plan for Energy Conservation and Carbon Dioxide Reduction sets specific targets for industries including steel, petrochemicals, and cement;[c] (iv) appliance standards—since 2005, the PRC's appliance energy efficiency standards and labeling programs have covered 37 types of products and more than 1.9 million product models, resulting in more than 500 terawatt-hours of electricity savings; and (v) the Energy Efficiency Leaders Program, which recognizes top energy-efficient products in different categories.[d]

---

[a] R. Bocca. 2024. China Strengthens 2024 Energy Efficiency Target, and Other Top Energy Stories to Read This Month. *World Economic Forum*. 25 March.

[b] Oxford Energy. 2024. Chinese Climate Policy: Energy Efficiency.

[c] *Climate Cooperation China*. 2024. China Issues Action Plan for Energy Saving and Carbon Reduction 2024–2025. 5 July.

[d] D. Sandalow et al. 2022. *Guide to Chinese Climate Policy 2022*. The Oxford Institute for Energy Studies.

[e] J. S. Lee. 2021. Greater Energy Efficiency Could Double China's Economy Sustainably. *International Energy Agency*. 15 February.

---

[56] Government of India, Ministry of Power. Green Energy Corridors; G. Hauber. 2024. Vietnam's Direct Power Purchase Agreement (DPPA) Decree Could Catalyze a New Era for Renewable Energy. *Institute for Energy Economics and Financial Analysis*. 29 July; and UNCTAD. 2023. *China's Policy Strategies for Green Low Carbon Development: Perspective from South–South Cooperation*.

[57] IEA. Policies Database. China Policies on Renewable Energy and Energy Efficiency (accessed 23 October 2024); and IEA. Policies Database. India Policies on Renewable Energy and Energy Efficiency (accessed 23 October 2024).

Developing Asian countries must build institutional capacity to support growth and enable the energy transition. Assessed countries must invest in their government's capacity to strengthen energy governance and enforce energy transition policies. Their regulatory quality scores and government effectiveness scores fall below the advanced economy and world medians, indicating vast scope for improvement as well as challenges in policy implementation. However, higher than world median statistical performance scores indicate that many developing Asian countries have statistical solid foundations for evidence-based policymaking in the energy sector.

Many assessed economies struggle to use data to design effective energy policies despite scoring high in statistical performance relative to the world, though they are 16% below advanced economies. Digitalizing and effective data management strategies can help governments design more effective policies. Quality and transparent data can help stakeholders and investors improve their understanding of the investment environment and attract needed capital for the energy transition. For example, India has improved its power sector data transparency through the National Power Portal, providing daily generation data across regions and fuel types, which enhances data-driven decision-making and strengthens the execution of clean energy policies.[58]

---

[58]   EMBER. 2023. *Asia Data Transparency Report 2023*.

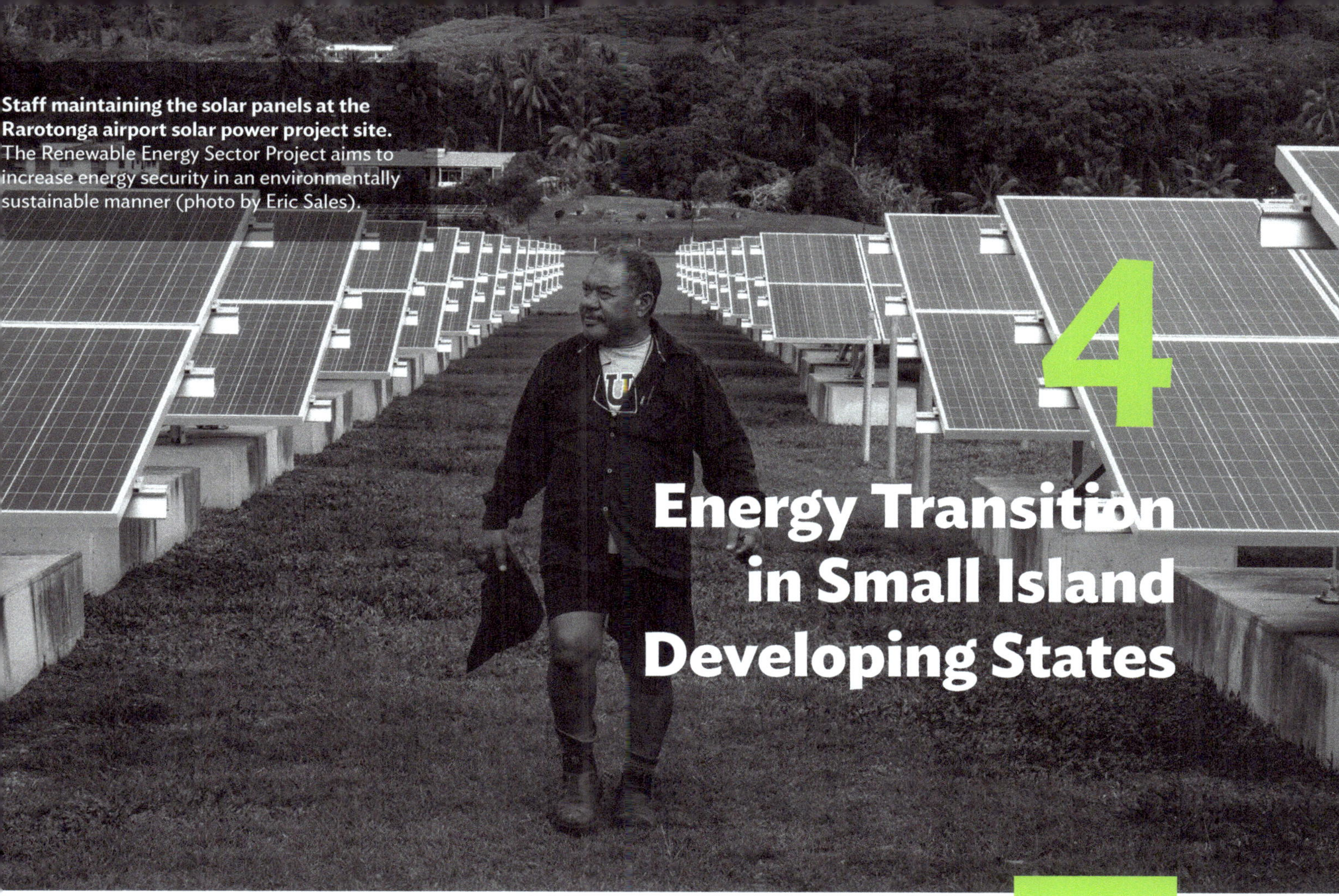

Energy Transition in Small Island Developing States

Staff maintaining the solar panels at the Rarotonga airport solar power project site. The Renewable Energy Sector Project aims to increase energy security in an environmentally sustainable manner (photo by Eric Sales).

Fossil fuel dependence, economic constraints, and exposure to climate risks underpin small island developing states (SIDS) energy transition.[59] These countries heavily rely on imported fossil fuels, making them susceptible to global price fluctuations. While their contribution to achieve global net-zero targets may be limited, they are also expected to face the hardest impacts of climate change.

Excluding Papua New Guinea, SIDS spend more than $1 billion annually on fossil fuels, representing nearly 80% of their total energy costs. Their net fuel imports as a share of total GDP are, on average, 6.8%.[60] Oil imports account for an average of 17.7% of merchandise imports[61] and 7.5% of GDP.[62] As a result, SIDS are vulnerable to energy price volatility, which has effects that ripple through all segments of their economies. As prices increase, so does the cost of electricity, water supplies, and other essential services.

---

[59] For the purpose of this chapter, SIDS include the Cook Islands, Fiji, Kiribati, Maldives, the Marshall Islands, the Federated States of Micronesia, Nauru, Niue, Palau, Papua New Guinea (PNG), Samoa, Solomon Islands, Timor-Leste, Tonga, Tuvalu, and Vanuatu.

[60] Data on net fuel imports is available for nine SIDS. Maldives has the highest dependency with 11.9% of GDP. PNG is an exception as it is a net fuel exporter with net fuel exports being 19.6% of total GDP.

[61] Calculated with data from the Observatory of Economic Complexity (https://oec.world/en) for the Cook Islands, Fiji, Kiribati, the Federated States of Micronesia, Nauru, New Caledonia, Niue, Palau, PNG, Samoa, Solomon Islands, Timor-Leste, Tonga, Tuvalu, and Vanuatu for the year 2020 (accessed 16 April 2024).

[62] Calculated with data from the Pacific Data Hub (https://pacificdata.org/) for the Cook Islands, Fiji, Kiribati, the Federated States of Micronesia, Nauru, Niue, Palau, PNG, Samoa, Solomon Islands, Tonga, Tuvalu, and Vanuatu (accessed 16 April 2024).

These increases eventually trickle down to consumer goods, affecting the ability of government, businesses, and households to spend, save, and invest. Climbing fossil fuel prices can also lead to increasing trade deficits and government spending, which can cause increased debt, currency depreciation, and borrowing costs.

Energy affordability varies widely within the SIDS, driven by income levels, tariff levels, tariff-setting methods, and subsidies for vulnerable customers. The average residential tariff in SIDS in 2019 was about $0.36 per kilowatt-hour (kWh), ranging from $0.16/kWh (Fiji) to $0.72/kWh (Solomon Islands and Vanuatu).[63] Lifeline tariffs are common but may sometimes be poorly targeted, benefiting wealthier consumers and putting additional pressure on government budgets.

Access to electricity in SIDS ranges widely, from 21% to 100%, with an average of 90% (the same as the world median), while access to clean cooking—which averages 50%—is significantly lower than the world median of 71%.[64] In some rural and remote areas, access to electricity continues to be relatively low. In those rural areas where access is relatively high, the productive uses are largely focused on tourism.

Grid reliability remains a persistent challenge in some SIDS due to inadequate maintenance, underinvestment, and susceptibility to disasters. Based on data from the Pacific Power Association, distribution losses average 18.0%, worsening energy affordability and reliability.[65]

Moreover, SIDS' energy infrastructure tends to be underdeveloped and highly vulnerable to climate-related disasters. The vulnerability of SIDS to climate change shocks is also well-documented and has important implications for infrastructure resilience. According to global indexes, SIDS are among the most vulnerable to climate-related risks (Figure 19). This vulnerability extends to their energy infrastructure, often exposed to extreme weather events, sea-level rise, and flooding.

The adoption of clean and renewable energy technologies is critical for SIDS, but multiple barriers must be addressed. Increased use of renewable energy can help reduce exposure to fossil fuel price volatility and reduce the drain of foreign exchange reserves. For instance, where solar energy has been introduced, costs have been considerably lower than those of diesel generation. Renewables (especially solar) remain vastly underutilized relative to the resource potential. However, uptake of renewable energy capacity in SIDS has been limited and ranges from 1% to 54%, with an average of 20%, compared with 29.1% for all developing Asia and the global average of 42%.[66]

SIDS face unique challenges in decarbonizing and advancing the energy transition. Land-use rights are often complex because of traditional ownership systems and limited available land, which can hinder the development of utility-scale renewable projects.[67] The remote nature of these islands further complicates logistics for maintaining and upgrading power systems, making it difficult to integrate renewable energy at scale. Moreover, the remoteness of these islands poses challenges for the disposal or recycling of renewable energy equipment at the end of its useful life.

63   S. Ugarte and A. Soakai. 2022. *Study on Pacific Clean Energy Financing Potential.* Pacific Region Infrastructure Facility.

64   IEA, International Renewable Energy Agency, United Nations Statistics Division, World Bank, and World Health Organization. 2023. *Tracking SDG7: The Energy Progress Report 2023.* World Bank.

65   Not all utilities in the 13 countries report data to the Pacific Power Association. Pacific Power Association. 2022. *Pacific Power Utilities: Benchmarking Report—2021 Fiscal Year.*

66   Ember. Yearly Electricity Data (accessed 6 September 2024).

67   ADB Private Sector Development Initiative. 2024. *Powering the Pacific: The Cost Implications of Renewable Energy.*

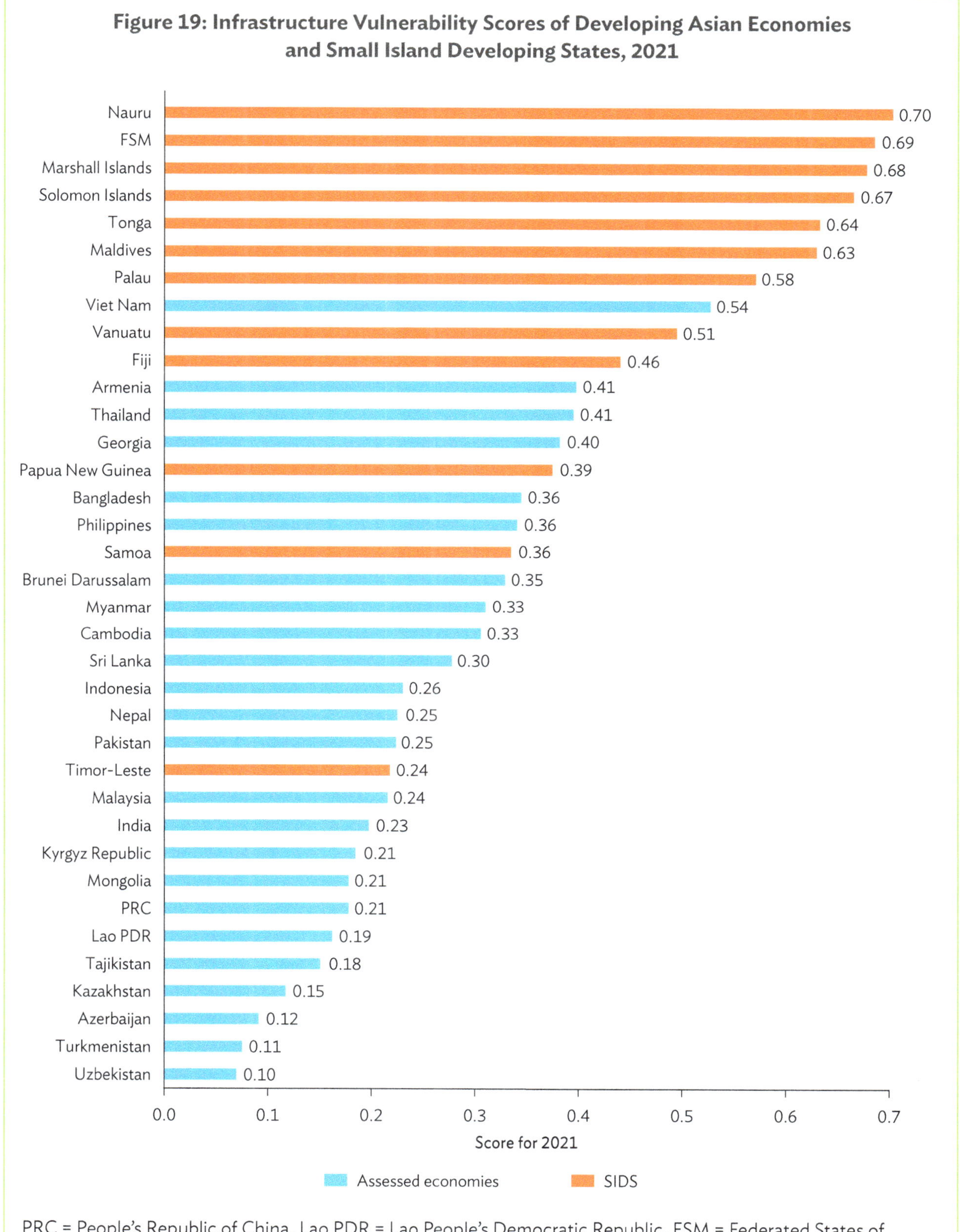

PRC = People's Republic of China, Lao PDR = Lao People's Democratic Republic, FSM = Federated States of Micronesia, SIDS = small island developing states.

Source: ND-GAIN data from IMF Climate Change Indicators Dashboard (accessed 2 July 2024).

The diffusion and production of clean energy technologies are further limited by SIDS' minimal manufacturing supply chains and technology ecosystem. This is borne by the United Nations Frontier Technologies Readiness Index, which includes seven SIDS and assesses national preparedness to use, adopt, and adapt frontier technologies equitably. The index combines indicators for information and communication technology, skills, R&D, industrial capacity, and finance, where SIDS score below the world median among these indicators (footnote 30). Energy transition efforts of many SIDS have been impeded by emigration of skilled labor to Australia, New Zealand, and elsewhere, making it difficult to find qualified professionals for utility operation and regulation. Fulfilling local expertise through targeted training programs and knowledge-sharing initiatives is crucial for addressing the lack of capacity and experience with renewables and driving energy transition.

Many islands also lack the financial resources to invest in sustainable and resilient energy infrastructure. With retail electricity tariffs often below the cost of service, utilities struggle to fund system improvements.[68] Electric utilities in SIDS are historically publicly owned and operated. Most SIDS have been slow to introduce independent power producers because of concerns about affordability, reliability, and a lack of proper public–private partnership frameworks to foster competitive tenders and to deal with unsolicited proposals. Domestic financing for renewables is also extremely limited in most countries, with many energy investments being funded with concessional loans or grants from international development partners.[69] These countries will need significant support to substantially scale up finance needed to set up climate-resilient renewable energy capacity and electricity infrastructure and address the unique challenges they face due to climate change.

Lastly, although all SIDS have submitted nationally determined contributions (NDCs), the regulatory environment in these countries needs to be strengthened to enable them to achieve their NDC targets. SIDS—most affected by climate change—have set ambitious decarbonization targets. Country commitment to climate action is a composite regulatory environment indicator that includes a country's commitment to its NDCs (Figure 17).[70] All SIDS have prepared and submitted NDCs— including renewable energy targets—to the United Nations Framework Convention on Climate Change.[71] Most have a plan, policy, strategy, or road map to increase the share of renewable energy generation in their territories, the only exceptions being Kiribati and Nauru.[72] Some have also provided incentives or mandates for clean energy and have worked to create adequate utility and regulatory capacity to scale up clean energy. However, SIDS will continue to need technical assistance in the planning, design, procurement, regulation, and operation of renewable energy facilities.

Improving and harmonizing regulatory frameworks is essential to creating a more attractive environment for private investment in clean energy and optimizing the use of public funds. Complex land use issues must be addressed for renewable energy development, and clear and stable policies must be created.

---

[68] Footnote 30, pp. 13–20.

[69] International Renewable Energy Agency. 2024. *Small Island Developing States at a Crossroads: The Socio-Economics of Transitioning to Renewables*.

[70] This has four main components: net-zero target commitment, sector policies or targets, coverage of GHGs under the NDCs, and net-zero target year.

[71] Climate Watch. NDC Tracker (accessed 11 August 2024).

[72] National Association of Regulatory Utility Commissioners. 2023. *Energy Regulatory Survey and Assessment Report for the Pacific Islands*.

Seven SIDS have formalized energy sector regulatory bodies; others regulate by way of direct control of a government line ministry and/or ministry of finance.[73] These regulators are often multisector, covering electricity and water and, in some cases, other sectors considered natural monopolies or strategically important. Formal regulatory bodies may have limited independence. Three of the seven formalized energy regulators have complete independence from other government entities (Fiji, Tonga, and Vanuatu), while the remaining four are only partially independent. The most common regulatory instruments to incentivize renewable energy integration in the region include adjustments to the grid code to facilitate the integration of new intermittent power generation and renewable energy licensing. Other instruments, such as differentiated tariffs based on technology and plant size, small-scale renewable energy policies, net metering, feed-in tariffs, emissions trading, and carbon pricing, are only applied by a few countries.

SIDS often require special approaches to effectively address their energy transition. These islands face unique challenges such as high dependency on imported fossil fuels, limited financial resources, and vulnerability to climate change impacts.[74] Grants and philanthropic initiatives can provide crucial support in several ways: (i) grants and concessional financing can help SIDS invest in renewable energy infrastructure like solar, wind, and hydro power, reducing their reliance on imported fuels; (ii) capacity building, training, and education programs can enhance local expertise in managing and maintaining renewable energy systems; (iii) technical assistance on integrating renewable energy into existing grids and developing decentralized energy solutions; and (iv) assistance in creating favorable policies and regulatory frameworks to attract private investment and ensure sustainable energy practices. Corporate social responsibility initiatives such as under the Moana Taka Partnership, under which Swire Shipping enables private sector companies and governments, to move recyclable waste out of the Pacific Islands to countries with competent and sustainable recycling plants, ensuring that waste products are properly recycled and processed.[75]

---

[73] Footnote 72, pp. 14–24.

[74] K. U. Shah and J. R. Biden Jr. 2022. Renewables and Energy Transitions in Small Island States. *International Institute for Sustainable Development*. 8 June.

[75] Secretariat of the Pacific Regional Environment Programme. 2020. *Moana Taka Partnership: A Guide for Pacific Island Countries & Territories*.

APPENDIX

ENERGY TRANSITION READINESS ASSESSMENT METHODOLOGY

Introduction

The energy transition readiness assessment (ETRA) framework provides a structured approach to evaluate and analyze the readiness of the Asian Development Bank (ADB) developing member countries (DMCs) for energy transition and developing Asian countries more broadly. "Readiness" refers to a country's capacity and preparedness to create an equitable, sustainable, and secure energy system that creates value for society and delivers on net-zero ambitions. It does so by using a multidimensional capability approach. This means that "readiness" is not measured against fixed, universal benchmarks but considers each country's unique circumstances, capacities, and needs. This approach allows for a tailored assessment that respects the individual pathways each country must take toward a clean energy future, ensuring that the strategies are both relevant and effective in their specific contexts.

The ETRA focuses on 25 out of the 41 ADB DMCs and Brunei Darussalam (collectively termed ETRA countries). There are 16 small island developing states covered separately as their data coverage—based on the choice of indicators—is insufficient to conduct the assessment. ETRA analysis is important because

(i)   The Asia and Pacific region is crucial for global efforts to reduce carbon emissions to net zero.

(ii)   ADB DMCs comprise a large part of the world's population and economic output.

(iii)   These countries are expected to grow quickly, increasing their energy needs.

(iv)   Most ADB DMCs import more energy than they export and are energy-intensive compared to the size of their economies.

(v)   These countries are also vulnerable to climate change-induced shocks.

(vi)   These factors make it urgent and important for the region to switch to sustainable energy systems.

Energy Transition Readiness Assessment Country Coverage

Full assessment is provided for		Partial assessment is provided for
Afghanistan	Myanmar	Cook Islands
Armenia	Nepal	Fiji
Azerbaijan	Pakistan	Kiribati
Bangladesh	Philippines	Maldives
Bhutan	Sri Lanka	Marshall Islands
Brunei Darussalam	Tajikistan	Micronesia, Federated States of
Cambodia	Thailand	Nauru
China, People's Republic of	Turkmenistan	Niue
Georgia	Uzbekistan	Palau
India	Viet Nam	Papua New Guinea
Indonesia		Samoa
Kazakhstan		Solomon Islands
Kyrgyz Republic		Timor-Leste
Lao People's Democratic Republic		Tonga
Malaysia		Tuvalu
Mongolia		Vanuatu

ETRA = energy transition readiness assessment.

Notes: Full assessment includes analysis based on the ETRA's 61 indicators and 7 dimensions. Partially assessed countries are not covered in this analysis but are included in Chapter 4: Focus on Small Island Developing States.

Source: ADB Classification.

The ETRA framework leverages the Energy Transition Index from the World Economic Forum and customizes it specifically for ADB DMCs. It delves into detailed indicators relevant to these countries, ensuring comprehensive coverage of all issues related to energy transition.

This approach identifies where countries stand in their energy transition journey and emphasizes their pathways toward sustainable energy systems. By focusing on each country's unique context and challenges, the framework provides a nuanced understanding of their readiness for energy transition. Instead of simply comparing countries against each other, the ETRA framework evaluates whether countries are on the path toward energy transition, recognizing that direct comparisons may not be as meaningful or constructive.

This document outlines the high-level methodology behind the creation and application of the ETRA framework, detailing the steps taken to develop a robust and contextually relevant assessment tool for energy transition readiness in the Asia and Pacific region.

Energy Transition Readiness Assessment Framework and Composition Overview

The ETRA framework is designed to assess countries based on their readiness for energy transition, encompassing economic, technical, and risk-related factors. It includes a comprehensive set of indicators and dimensions reflecting the multifaceted nature of energy transition readiness.

Countries are scored on a scale of 0 (worst) to 100 (best) across seven dimensions, composed of 61 indicators that are further categorized into subdimensions that help capture specific focus areas.

For detailed indicator descriptions and sources, please see the ETRA Codebook.

Energy System: Assesses a country's capability to ensure energy security and transition away from fossil fuels at speed and scale. Scaling up low-carbon energy sources while transitioning from fossil fuels depends on the extent of current carbon lock-in, resource endowments, and system architecture. This dimension measures the ability of countries to diversify energy imports and supply, integrate clean energy sources, improve emission and energy intensities, and reduce fossil fuel dependence.

Energy Economy Linkages: Assesses a country's capability to decouple energy use from economic growth and development. Energy is interlinked with economic growth, impacting industrial production, export revenue, and fiscal stability. This dimension assesses the ability of countries to adopt best-in-class energy efficient technologies to improve both the emission and energy intensities of its economy, increase trade in low-carbon technology products, and reduce the carbon footprint of its exports.

Infrastructure System: Assesses a country's capability to integrate low-carbon energy sources, optimize energy demand, and improve transportation systems and development of information and communication technology. It measures countries' climate resilience and coping capacity, energy infrastructure quality, and physical and digital infrastructure.

Social System: Assesses a country's capability to provide affordable energy and services and to recognize and address the distributional impacts of the energy transition. This dimension considers select key parameters to assess countries' capabilities to ensure access to affordable energy and energy services, rationalized electricity tariffs for vulnerable sections of the population, and institutional mechanisms to identify and address distributional implications of the energy transition on vulnerable stakeholder groups.

Technology and Diffusion Ecosystem: Assesses a country's capability to innovate, transfer, produce, and deploy new energy technologies. It measures a country's research and development ecosystem, availability of skilled human capital, clean energy technology diffusion, and comparative advantage.

Macroeconomic and Investment Environment: Assesses a country's capability to attract diverse sources of domestic and foreign capital and its macroeconomic fundamentals. It focuses on monetary and fiscal conditions, financial stability and risk, openness to investors, and clean energy investment flows that influence and reflect its ability to attract and allocate funding for the energy transition.

Regulatory Environment: Assesses a country's ability to create stable, comprehensive energy transition policies. Effective regulations align stakeholder interests, enhance governance, support modern infrastructure and innovation, ensure energy security and resilience, guide sustainable investments, minimize macroeconomic risks, and promote social equity by ensuring affordable energy and a just transition.

Energy Transition Readiness Assessment Dimensions, Subdimensions, and Indicators

 **Energy System**

Energy Security

- Diversity of Electricity Generation
- Diversity in Energy Supply
- Diversification of Energy Imports
- Net Energy Imports

Clean Electrification

- Share of electricity in final energy consumption
- Share of clean energy sources in electricity generation

Fossil Fuel Dependence

- Average Life of Coal Plants
- CO_2 Intensity of TPES
- CH_4 Intensity of Production

 **Energy Economy Linkages**

Energy-Emissions-Economy Nexus

- Energy Intensity
- Emissions Intensity

Energy Trade

- Emissions Intensity of Exports
- Trade Balance in LCT

Resource Dependence

- Net Fuel Imports
- Fossil Fuel Rents
- Energy Subsidies

 **Infrastructure System**

Climate Resilience and Coping Capacity

- Water Stress
- Lack of Coping Capacity
- Infrastructure Vulnerability

Energy Infrastructure Quality

- Transmission and Distribution Losses
- Flexibility in Electricity System

Physical and Digital Infrastructure

- Quality of Infrastructure
- Digital Infrastructure Readiness

 **Social System**

Energy and Transportation Access

- Per Capita Energy Consumption
- Rural Electricity Access
- Clean Cooking and Heating Access
- Access to Public Transportation

Energy Affordability

- Electricity Prices for Industry
- Household Electricity Prices
- Petrol Prices

Social Equity and Development

- Human Development Index
- Social Protection Cover
- Labor Rights

Environmental Health and Climate Vulnerability

- PM2.5 Air Pollution, Mean Annual Exposure (micrograms per cubic meter)
- Vulnerability to Climate Disruptions

 **Technology and Diffusion Ecosystem**

Clean Energy Technology Development and Innovation Capacity

- Public R&D
- Frontier Technologies Readiness

Human Capital

- Talent Competitiveness
- Jobs in Low Carbon Industries

Technology Diffusion

- Renewable Capacity Build-out
- Comparative Advantage in LCT

 **Macroeconomic and Investment Environment**

Monetary and Fiscal Indicators

- General Government Debt
- Fiscal Balance
- Consumer Prices Inflation
- Exchange Rate Depreciation
- Real Interest Rates

Financial Stability and Risk

- Domestic Credit to Private Sector
- NPA Share of Gross Bank Loans
- Risk of Banking System Failure
- Sovereign Credit Rating

Investment Climate

- Investment in Clean Energy
- FDI in Clean Energy Sector
- De Jure Financial Globalization

Regulatory Environment

Governance Quality

- Regulatory Quality
- Government Effectiveness
- Statistical Performance

Energy Policy Framework

- Renewable Energy Policies
- Clean Cooking Policies
- Energy Efficiency Policies
- Electricity Access Policies

Climate Commitments

- Climate Action Commitment

CH_4 = methane, CO_2 = carbon dioxide, FDI = foreign direct investment, LCT = low-carbon technology, NDC = nationally determined contribution, NPA = nonperforming asset, PM2.5 = particulate matter, R&D = research and development, TPES = total primary energy supply.
Source: Asian Development Bank.

Energy Transition Readiness Assessment Computation and Scoring

Dimension Scoring

Country performance for each dimension was based on a score of 0 (worst) to 100 (best). Each dimension score was based on the weighted average score of its constituent indicators.

Weighting was determined as follows:

- Each subdimension within a dimension was given equal weight.
- Within each subdimension, indicators were given equal weight.
- The overall weight for an indicator was calculated as:

$$w_i = \frac{1}{N_{sd}} \cdot \frac{1}{N_i}$$

where

w_i = weight assigned to the indicator

N_{sd} = number of subdimensions in a dimension

N_i = number of indicators within a subdimension

Note that subdimensions were not scored and were used only for categorization and weighting purposes.

For each dimension, country, and year:

(i)   The weighted average of the indicator scores within that dimension was calculated.

(ii) Only indicators with available data (non-"n/a" scores) were included in this calculation.

(iii) The products of each indicator's score and its weight were summed.

(iv) This sum was divided by the total weight of available indicators to get the final dimension score.

A dimension score was calculated only if the sum of weights for available indicators was at least two-thirds of the total possible weight for that dimension. If this threshold was not met, the dimension score was recorded as "n/a."

Indicator Scoring

To enable standardized and comparable measures of performance across indicators and to allow the calculation dimension scores, each indicator's underlying data was normalized on a scale of 0 (worst) to 100 (best) using the following formula:

$$100 \times \frac{C - Min_s}{Max_s - Min_s}$$

where

C = country indicator value

Min_s = sample minimum

Max_s = sample maximum

The sample minimum and sample maximum are the lowest and highest values of the overall indicator dataset.

Indicators for which a higher value indicates a worse outcome (see the ETRA Codebook for details) rely on a normalization formula that, in addition to converting the series to a 0-to-100 scale, reverses the score, so that 0 and 100 still correspond to the worst and best:

$$100 - 100 \times \frac{C - \text{Min}_s}{\text{Max}_s - \text{Min}_s}$$

Note that the sample sizes vary indicator-by-indicator depending on the availability of the data in the underlying sources.

However, many indicator datasets contained outlier values. In these situations, an outlier handling process was applied. This process assigned scores of 0 to low outliers (values significantly lower than the majority of the data) and scores of 100 to high outliers (values significantly higher than the majority of the data).

Dealing with Outliers

Outlier detection was performed on the combined datasets to identify unusual or extreme values that may impact the scoring process.

Based on the indicator-by-indicator assessment, the indicators were marked where outlier detection was not required. Indicators based on indexes or standardized scoring were kept out of the outlier detection process as they would have already gone through such filtering during their creation (see the ETRA Codebook for details).

The Winsorization method was used to handle the outliers.[1] Winsorization is a method for handling outliers by adjusting extreme values, bringing them closer to the central data distribution rather than removing them. This technique is used when indicators exhibit high skewness or kurtosis, specifically when skewness values exceed ±2 and kurtosis values exceed 3.5. Winsorization replaces outliers with values closer to the next nearest observations within acceptable limits. This transformation softens the impact of extreme values, reducing skewness and "tailedness" (kurtosis) in the dataset while maintaining its overall structure. This transformed data is then used for the scoring process.

Missing Value Imputation

The imputation strategy aims to fill data gaps using the most recent available information within a specified timeframe. The process was as follows:

(i)   Data were examined for each country indicator pair from 2023 backward to 2005.

(ii)   When encountering a missing value for a given year, the ETRA looked back up to 5 years to find the most recent available data point.

(iii)   If a valid value was found within this 5-year window, it was used to impute the missing value.

---

[1]   European Union. 2023. Joint Research Centre Training on Outlier Detection.

Use of Averages for Data Smoothing

To smooth out short-term fluctuations and highlight longer-term trends, an averaging technique was applied to specific indicators before scoring:

(i)   A mapping sheet was used to identify which indicators required averaging.

(ii)   For designated indicators, a 3-year moving average was applied (see the ETRA Codebook for details):

    (a)   For each year from 2023 back to 2010, the average of the current year and the 2 preceding years was calculated.

    (b)   Only valid (non-null) values were used in this calculation.

    (c)   In cases where fewer than 3 years of data were available, the available data for the average was used.

(iii) Some indicators were treated differently:

    (a)   Those flagged for 1-year average retained their original values.

    (b)   Indicators not flagged for averaging remained unchanged.

This helped mitigate the impact of year-to-year volatility, providing a more stable basis for assessing long-term trends in energy transition readiness.

It presented a complete picture of energy transition readiness across diverse countries and periods while maintaining methodological rigor and transparency.

Indicator Selection, Sourcing, and Data Validation

Selection of Indicators

(i)   Purpose and Approach

    ☐   The selection of indicators is foundational to the ETRA framework as it determines the breadth and depth of the assessment. Indicators were chosen to comprehensively represent the economic, technical, policy, social, and risk-related aspects of energy transition readiness.

(ii)   Criteria for Selection

    ☐   Relevance: Indicators must be directly related to the dimensions of the ETRA framework and provide meaningful insights into the readiness for energy transition.

    ☐   Reliability: Data for the indicators must be accurate, consistent, and sourced from credible institutions.

    ☐   Availability: Indicators should have data available across a wide range of countries and over multiple years to enable comprehensive and comparative analysis.

Data Sources and Validation

To ensure the reliability of the framework, data was sourced from highly reputable organizations and databases such as ADB, the International Energy Agency, the International Monetary Fund, the United Nations, and the World Bank.

Data underwent rigorous validation checks to ensure accuracy and consistency. This included cross-referencing with multiple sources and applying statistical validation techniques.

REFERENCES

Asia-Pacific Economic Cooperation. 2022. *APEC Energy Demand and Supply Outlook*. 8th ed. Vol. 1. Asia Pacific Energy Research Centre.

Asian Development Bank (ADB). 2012. *Climate Risks and Adaptation in the Power Sector*.

——. 2023a. *Asia in the Global Transition to Net Zero: Asian Development Outlook 2023 Thematic Report*.

——. 2023b. *Asia-Pacific Climate Report 2024 Catalyzing Finance and Policy Solutions*.

——. 2023c. New Agreement Aims to Retire Indonesia 660-MW Coal Plant Almost 7 Years Early.

——. 2023d. Road Map to Scale Up Solar Irrigation Pumps in Bangladesh (2023–2031).

——. 2024. Building Resilient and Responsible Critical Minerals Supply Chains for the Clean Energy Transition. ADB Brief No. 298.

ADB, Bloomberg Philanthropies, ClimateWorks, and Sustainable Energy for All. 2023. *Renewable Energy Manufacturing: Opportunities for Southeast Asia*.

ADB Private Sector Development Initiative. 2024. Powering the Pacific: The Cost Implications of Renewable Energy.

ASEAN Centre for Energy. 2023a. A Study of Energy Savings with Focus on Demand Side Management in ASEAN.

——. 2023b. ASEAN Energy Ministers Statement on Power Grid Development.

——. 2024. Pushing Indonesia's Decarbonisation Agenda Through Low-Carbon Hydrogen National Strategy.

Asian Power. 2024. Asia to Lead Grid Expansion Investments. 23 February.

Bain & Company. 2023. Voluntary Carbon Markets in 2023: A Bumpy Road Behind, Crossroads Ahead.

Bernama Energy. 2024a. Power smart meters in 9.1 million households by 2026. 27 September.

——. 2024b. Smart Meter Technology Transforms National Energy Management. 3 August.

Bloomberg. 2023. New Energy Outlook 2022: Global Net Zero Will Require $21 Trillion Investment in Power Grids. Blog. 2 March.

BloombergNEF. 2024. Energy Transition Investments (accessed 8 July 2024).

Carbon Market Institute. 2024. *International Carbon Market Update States and Trends in the Asia Pacific*.

China Briefing. 2024. Understanding the Relaunched China Certified Emission Reduction (CCER) Program: Potential Opportunities for Foreign Companies.

Climate Watch. 2024. Climate Watch Data—Compare All Targets (accessed July 2024).

Collaborative Labelling and Appliance Standards Program (CLASP). 2023. *China's MEPS Lead to Major AC Market Transformation.*

Ember. 2024. Yearly Electricity Data (accessed 6 September 2024).

Emissions Database for Global Atmospheric Research (EDGAR). GHG Emissions Database (accessed 11 July 2024).

Enerdata. 2024. World Energy and Climate Statistics 2024.

The Energy Progress Report. https://trackingsdg7.esmap.org/ (accessed 23 December 2024).

Energy Sector Management Assistance Program. 2016. Indonesia Clean Cooking: ESMAP Supports Innovative Approaches to Build the Local Cookstoves Market, Helps Increase Access. 2 September.

———. 2022. RISE Report 2022.

Energy Watch. 2022. The Go-to-Guide for Malaysia's Smart Meter. 23 December.

European Union. 2023. Joint Research Centre Training on Outlier Detection.

Fortune Business Insights. 2024. Asia Pacific Commercial Air Conditioner Market Size [2029].

Glasgow Financial Alliance for Net Zero. 2023. Financing the Managed Phaseout of Coal Fired Power Plants in Asia Pacific.

Global Energy Monitor. Global Coal Plant Tracker.

Global Methane Pledge. 2023. 22 November.

Government of India, Ministry of Power. Green Energy Corridors.

Government of the People's Republic of China, State Council. 2024. China leads global green patent applications with over 50 pct share. News release. 7 August.

Government of Singapore, Ministry of Sustainability and Environment. 2023. Singapore Sets Out Eligibility Criteria for International Carbon Credits Under the Carbon Tax Regime.

Hauber, G. 2024. Viet Nam DPPA. Institute for Energy Economics and Financial Analysis. 29 July.

Hiep, L. H. 2024. The Unexpected Twist in Vietnam's Renewable Energy Saga. Fulcrum. 4 January.

Hydrogen Council. 2023. Hydrogen Insights 2023.

International Energy Agency (IEA). 2018. The Future of Cooling – Opportunities for energy-efficient air conditioning.

———. 2021. Average age of existing coal power plants in selected regions in 2020.

———. 2022a. *Roadmap Towards Sustainable and Energy-Efficient Space Cooling in the Association of Southeast Asian Nations.*

———. 2022b. *World Energy Outlook 2022: Global Net Zero Will Require $21 Trillion Investment in Power Grids.*

———. 2022c. Skills Development and Inclusivity for Clean Energy Transitions.

———. 2023a. Net Zero Roadmap: A Global Pathway to Keep the 1.5C Goal in Reach—2023 Update.

———. 2023b. Space Cooling.

———. 2023c. *Clean Energy Supply Chains Vulnerabilities.*

———. 2023d. *Electrolysers.*

———. 2024a. *World Energy Investment 2024.*

———. 2024b. Cost of Capital Observatory.

———. *Energy Systems—Buildings.*

———. *SDG7: Data and Projections—Access to Clean Cooking.*

———. *SDG7: Data and Projections—Access to Electricity.*

———. *IEA Tracking Clean Energy Innovation Focus on China.*

———. *ASEAN Renewables: Investment Opportunities and Challenges.*

———. *Scaling Up Private Finance for Clean Energy in Emerging and Developing Economies.*

———. *Reducing the Cost of Capital.*

———. Global Green Hydrogen Review.

IEA, Policies Database. China Policies on Renewable Energy and Energy Efficiency (accessed 23 October 2024).

———. India Policies on Renewable Energy and Energy Efficiency (accessed 23 October 2024).

Inter-Agency Standing Committee and European Commission. 2024. INFORM Report 2024: 10 Years of INFORM.

International Carbon Action Partnership (ICAP). 2024. India adopts regulations for planned compliance carbon market.

International Labour Organization. 2020. World Social Protection Data.

International Monetary Fund (IMF). 2024. *World Economic Outlook Database.*

International Renewable Energy Agency (IRENA). 2024. *Small Island Developing States at a Crossroads: The Socio-Economics of Transitioning to Renewables.*

Janssen, P. 2024. South-East Asia Poised for Sustainable Bond Lift-off. *The Banker.* 29 April.

Kauf, N. and K. Petrichenko. 2023. *Energy Efficiency Market Report 2022 and Implications for Southeast Asia.* International Energy Agency.

Kose, M. Ayhan et al. 2022. A Cross-Country Database of Fiscal Space. *Journal of International Money and Finance*. 128. November. 102682.

Meticulous Research. 2023. Asia-Pacific Air Conditioners Market – Opportunity Analysis and Industry Forecast (2023–2030).

National Association of Regulatory Utility Commissioners (NARUC). 2023. Energy Regulatory Survey and Assessment Report for the Pacific Islands.

Observatory of Economic Complexity. https://oec.world/en (accessed 16 April 2024).

Observer Research Foundation. 2022a. To Price or not to Price? Making a Case for a Carbon Pricing Mechanism for India.

————. 2022b. Demand–Supply Dynamics in Bangladesh's Energy Sector.

Ogbevoen, L. What is Digital Readiness? Oden.

Pacific Data Hub. https://pacificdata.org (accessed 16 April 2024).

Pacific Power Association. 2022. *Benchmarking Report: 2021 Fiscal Year*.

Petronas. 2024. PETRONAS and Partners to Advance Methane Emissions Reduction Efforts in Southeast Asia Region.

Rasmussen, T. N. and A. Roitman. 2011. Oil Shocks in a Global Perspective: Are They Really that Bad? IMF Working Paper: WP/11/194.

Reuters. 2023. Explainer: Japan's carbon pricing scheme being launched in April.

Rystad Energy. 2024a. *Asia Power Grid Investment Report*.

Scoccimarro, E. et al. 2023. Country-level energy demand for cooling has increased over the past two decades. *Communications Earth and Environment*. 4(1). p. 208.

SE for ALL. 2023. Chilling Prospects—Accession to Cooling Gaps 2023—Regional Risks.

Secretariat of the Pacific Regional Environment Programme (SPREP). 2020. *Moana Taka Partnership: A Guide for Pacific Island Countries & Territories*.

Senstar. Energy Sector Vulnerabilities.

Shah, K. U. and J. R. Biden, Jr. 2022. Renewables and Energy Transitions in Small Island States. *International Institute for Sustainable Development*. 8 June.

Sharma-Kushal, S. 2021. Why Should the Asian Development Bank Care about Sustainable Cooling? E3G.

Shaw, V. 2024. Longi Green Energy sets world record for solar module efficiency at 25.4%. *PV Magazine*. 24 October.

Sivak, M. 2009. Potential energy demand for cooling in the 50 largest metropolitan areas of the world: Implications for developing countries. *Energy Policy*. 37(4). pp. 1382–1384.

Smart Energy International. 2023. What is Power Quality and Why is it Important?

Ugarte, S. and A. Soakai. 2022. Study on Pacific Clean Energy Financing Potential. Pacific Region Infrastructure Facility.

UN DATA. UN Data Portal (accessed 18 July 2024).

UNCTAD. 2023. *Technology and Innovation Report 2023*.

———. 2024. *2024 World Investment Report*.

United Nations. Human Development Index (accessed 6 September 2024).

United Nations Climate Change. 2024. Action to Decarbonize High Emitting Industrial Sectors Gathers Pace. News article. 6 May.

US Department of State. 2023. US–Kazakhstan Joint Statement on Accelerating Methane Mitigation to Achieve the Global Methane Pledge. Press release.

US Energy Information Administration. International Energy Statistics (accessed 22 October 2024).

World Bank. World Development Indicators.

———. 2014. Indonesia: Government will provide universal access to clean cooking practices. Press release. 14 August.

———. 2020. Inflation, Consumer Prices (Annual %) (accessed July 2024).

———. 2022. Climate Investment Opportunities in India's Cooling Sector.

World Economic Forum. 2021. 5 Ways to Boost Renewable Energy Investment in Developing Economies.

———. 2024a. *Net-Zero Industry Tracker 2024*.

———. 2024b. *Transitioning Industrial Clusters*.

World Resources Institute Climate Watch. 2021. 2020 NDC Tracker.

———. 2023. 9 Things to Know About National Climate Plans (NDCs).

Xinhua. 2024. Economic Watch: China's clean energy sectors gain edge via innovation, supply chain. 30 April.